Quarks and Gluons

A Century of Particle Charges

Quarks and Gluons

A Century of Particle Charges

M Y Han
Duke University

World Scientific
Singapore • New Jersey • London • Hong Kong

Published by

World Scientific Publishing Co. Pte. Ltd.
P O Box 128, Farrer Road, Singapore 912805
USA office: Suite 1B, 1060 Main Street, River Edge, NJ 07661
UK office: 57 Shelton Street, Covent Garden, London WC2H 9HE

Library of Congress Cataloging-in-Publication Data
Han, M. Y.
    Quarks and gluons : a century of particle charges / M.Y. Han.
      p.   cm.
    ISBN 9810237049. -- ISBN 9810237456 (pbk.)
    1. Particles (Nuclear physics) -- History.   I. Title.
QC793.16.H36   1999
539.7'25--dc21   98-53246
                                                                             CIP

British Library Cataloguing-in-Publication Data
A catalogue record for this book is available from the British Library.

Copyright © 1999 by World Scientific Publishing Co. Pte. Ltd.

All rights reserved. This book, or parts thereof, may not be reproduced in any form or by any means, electronic or mechanical, including photocopying, recording or any information storage and retrieval system now known or to be invented, without written permission from the Publisher.

For photocopying of material in this volume, please pay a copying fee through the Copyright Clearance Center, Inc., 222 Rosewood Drive, Danvers, MA 01923, USA. In this case permission to photocopy is not required from the publisher.

Printed in Singapore.

Dedicated to Changki, Chris, Grace, Chris, Rita, Tony and last but not least, Leilani.

Acknowledgments

I thank with a deep sense of appreciation and friendship my editor, Mr. Andrew Robinson, for his care and concern in this project from beginning to end, and for burning much midnight oil in cyberspace. The east coast of the United States and Singapore being opposite sides of the globe, it was a truly global round-the-clock collaboration.

I would like to thank the staff of World Scientific, including Ms. Joy Marie Tan (editor), for their meticulous attention to the book and for their contributions to its improvement. Finally, I wish to express my gratitude to Dr. K K Phua, Chairman of World Scientific, and Dr. Da Hsuan Feng of Philadelphia, one of its editors, for their faith and encouragement in the project.

Contents

Prologue: A Century of Particle Charges — 1

1. The Electron: The Quantum of Electricity — 7
2. Mass: $E = mc^2$ and All That — 17
3. The Photon: No Charge, No Mass — 27
4. The Spin: If It's Round, It Rolls — 35
5. Antimatter: A Mirror Image — 45
6. The Nucleus: A Whole New Ball Game — 53
7. The Strong Force I: Nucleons — 65
8. The Weak Force: A Whisper in the Night — 75
9. The Strong Force II: Hadrons — 89
10. The Quark: The Queen of Fractions — 101
11. The Origin of Quarks and Gluons — 113

Epilogue: More Quarks, More Leptons and More Charges — 125

Appendix 1: Annotated Chronology 129

Appendix 2: Powers of Ten 143

Appendix 3: The Nobel Prizes in Physics 147

Prologue

A Century of Particle Charges

The history of the search for the origin of matter had its modern genesis about a hundred years ago. Two discoveries — that of the electron in 1897 and of the photon in 1900 — mark, more than any other events at the turn of the 20th century, the beginning of an epoch that has come to be called modern physics; the physics of atoms and beyond. Electrons and photons represent the smallest known objects in their respective worlds; the electron is the tiniest bit of matter, while the photon is the tiniest bit of radiation.

During the first three decades of the 20th century, a host of bold new concepts were introduced and groundbreaking new theories were formulated — the theories of relativity and quantum physics. The structure of atoms was laid bare and understood to an unprecedented level of accuracy. What provided the basis of our understanding of atoms during this period — what has come to be referred to as 'the 30 years that changed the world' — was our unraveling of the secrets of the ways in which electrons and photons interacted with each other, at the microscopic scale of the atomic structure.

What was not at first anticipated was just how fundamental a role 'the theory of electrons and photons' would come to play in subsequent research. From the 1930s onwards we began to explore beyond atoms and probe further into the world of the atomic nuclei — the world of violent reactions among the subatomic particles — and eventually to the world within protons and neutrons themselves. This was the world of the so-called quarks and gluons, the as yet unobserved layer of the origin of matter. The basic properties of the electron and photon — specifically the electric charge of the electron and the way in which it interacts with photons — form the basic framework from which we attempted to understand the goings-on within the reduced dimensions of the atomic nuclei and of quarks and gluons. The smaller the scale of the elementary particles, the less certain we became of the applicability of the rules of the world of atoms to that of the subnuclear particles. We sorted out the rules and principles that govern the world of electrons and photons, and zeroed in on those that we considered to be universal truths that would serve as guiding lights.

One such guiding principle — which would eventually turn out to be the most fundamental in the world of elementary particles — is based upon the concept of an intrinsic property of particles called the charge (the electric charge in the case of electrons) and the fact that these charges are known to obey a rule of absolute conservation — hence the name 'conserved charges.' No net amount of the electric charge can be created or destroyed, and consequently in any reactions among particles, the total amount of electric charges before and after must always balance out — a strict 'zero-sum' rule.

This aspect of the electric charge — that it is conserved — is not exactly something new to the 20th century; it has been known since the dawn of the age of electricity. What the discovery of the electron established was that the amount of the electric charge on a single electron served to define the smallest amount of electricity, hence defining the fundamental scale of the charges. The electron is not only the smallest known bit of matter, but also serves to define the quantum of electric charge. Whatever its ultimate physical origin may be, the electric charge is the very essence of the electron. As charges are transferred from particles to particles, they always occur in units of the electronic charge in accordance with the strict rule of conservation — there are no exceptions to this zero-sum rule. How little did we anticipate, back in the days of atomic physics, that this property of the electric charge would evolve into one of the most fundamental principles — that of the conserved charges — to guide us in our quest into the lesser-known world of atomic nuclei, and eventually of quarks and gluons.

By the early 1930s, the physics of the atoms were more or less completely understood. From that time on, the relentless search for the origin of matter has gone through several layers of ever-diminishing dimensions, probing deeper and deeper into the central core of atoms. Passing through the 'cloud' of electrons that form the outer 'skin' of atoms, we came face to face to the central core of an atom, a tight bundle of mass and electric charge called the atomic nucleus; within which we found the world of the nucleons — the generic name for protons and neutrons — a world the likes of which we had never seen before. It is a strange world of protons and neutrons — that is, the charged and uncharged nucleons — under the influence of two brand new forces; the strong and the weak nuclear forces. We could not have encountered these new forces before we were able to probe into the innards of atomic nuclei; the strong and the weak nuclear forces do not exist outside the shells of atomic nuclei.

Confronted by brand new forces of nature, confusing arrays of ever-increasing particles that these new forces are capable of producing, and all manner of strange reactions among all these new particles, we desperately looked for a way in which we could hope to make sense of it all. And it was here that we began to emulate the electric charge and its zero-sum rule. We expanded and extended the concept of the conserved charges into the domain

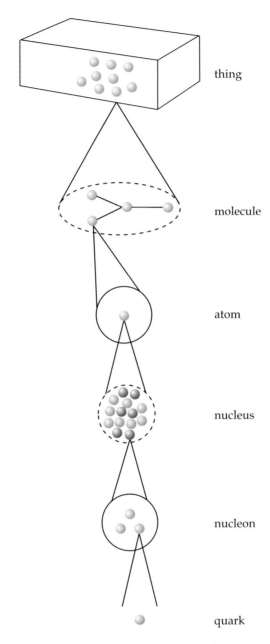

Figure 1 Layers of ever-diminishing dimensions.

of subnuclear particles by inventing several new breeds of conserved charges.

The first of such new conserved charges was that of the nucleonic charges. With a proper assignment of the nucleonic charges to nucleons — positive one unit of the nucleonic charge for nucleons and negative one unit of the nucleonic charge for antinucleons — a zero-sum rule for the nucleonic charges could be established: In all reactions involving nucleons, the net nucleonic charge would not change. The introduction of the nucleonic charge was only the beginning of what came to be a hallmark of elementary particle physics in the latter half of the 20th century; the invocation of a string of new conserved charges that would help us to make some sense out of what at first appeared to be the quite chaotic world of particles.

Nucleons soon turned out to have an assortment of related particles and quasiparticles — those that live only a fraction of a fraction of a fraction of a second. This extended family, dubbed baryons, necessitated the extension of the concept of the nucleonic charge to what is called the baryonic charge, and again with a proper assignment of the baryonic charges to the baryons it was possible to claim the baryonic zero-sum rule among particle reactions.

Not to be outdone, the electron turned out to have a sidekick of its own, namely the neutrino. Electrons and neutrinos were generically named leptons, and it soon became evident that leptons too could be endowed with a conserved charge all their own; the leptonic charge and its zero-sum rule. The concept of the conserved charge was by then extended to include the nucleonic, baryonic and leptonic charges, as well as the original electric charges.

Coming into the 1960s, we were able for the first time to probe deeper into yet another level of reduced dimensions; the realm within the nucleons themselves. The body of evidence — albeit all circumstantial to this date — began to point to the irrefutable conclusion that the nucleons that make up the atomic nuclei might very well have internal structures themselves. A brand new set of particles that made up protons and neutrons were called for. These are the quarks. Nucleons have a three-quark structure, much like a tritium nucleus or a helium-3 isotope nucleus that is made up of three nucleons (two protons and a neutron for the tritium nucleus and two neutrons and a proton for the helium-3 isotope nucleus). The existence of quarks necessitated a further emulation of the conserved electric charges — the quark charge.

As we look back at the development in the past one hundred years in the science of atoms, nuclei, protons and neutrons, and finally of quarks, we see that a tremendous amount of new knowledge has been uncovered concerning the ultimate origin of matter, not to mention the discovery of two great theories; relativity and quantum physics. Threading through all these developments of 20th century physics, however, there exists one unifying principle that transcends across the reduced dimensions of the subatomic and subnuclear worlds, and that is the concept of conserved charges — the nucleonic, baryonic, leptonic and quark charges. The concept of the conserved electric charge — that of electrons and photons — has been extended all the way to the world of conserved quark charges (also known as the color charges), those of quarks and gluons. In this book, I trace out the major developments in atomic, nuclear, particle and quark physics that span the past one hundred years, by following this particular path; the concept of conserved charges. In fact, the book might have been alternatively titled, 'a brief history of charges.'

1

The Electron: The Quantum of Electricity

The electron is perhaps the most remarkably unremarkable particle that we have come to know. It is unremarkable in the sense that it is so common; as one of the basic ingredients of atoms, it is one of the most abundant of all the elementary particles of nature. Electrons are the body and soul of everything under the Sun that is electric or electronic. They are everywhere, either by their unattached selves, flowing down a piece of wire, or whirling about within every single atom and molecule in the Universe. What is at the same time remarkable about this particle is that it takes several superlatives to characterize the little fella — it is the smallest, the lightest, the first subatomic particle to be discovered, and the oldest truly elementary particle, among others.

To say that electrons are indescribably minuscule bits of matter is an exercise in understatement, even by the standard of subatomic physics. One gram of water, a mere drop, contains some 30 trillion billion water molecules, each of which contains ten electrons apiece. Another example: when you switch on a pocket calculator, about 2.5 billion electrons rush through its circuitry every one microsecond. To put it perhaps more dramatically, one can say that the size of an electron is to the size of a person what the latter is to the size of the Milky Way galaxy. Mighty tiny indeed.

Of all subatomic particles that have mass, however minuscule, the electron is the smallest in size and the lightest in mass.

Another superlative about the electron is historical. The electron, discovered in 1897 by Joseph John Thomson (England, 1856-1940), is the first ever subatomic particle on record; it is the oldest (about a hundred years old) known particle that remains truly elementary. Often, particles first considered to be elementary — that is, without any internal structure — were later proven to be otherwise, revealing another layer of internal structures. But after all these years, despite all manner of 'abuses' at the hands of scientists — such as getting kicked and kicked hard until accelerating to near the speed of light, and then being hurled toward a head-on crash with another electron coming the other way also at near the speed of light — the electron has never yielded any sign of having any further internal structure; it is truly an elementary dot, complete in itself. Rightfully, the electron has the title of the oldest truly elementary particle.

The discovery of electrons as the ultimate carriers of electricity not only established the materialistic origin of electricity — that it was not some mysterious power emanating out of the terminals of a battery — but also opened up a whole new vista for our microscopic understanding of the flow of electricity. This knowledge would eventually lead, some half a century later, to the development of the so-called high technology and information age, based on such new devices as semiconductors, lasers, superconductors, and optoelectronics.

Reaching far beyond its immediate consequences for the nature of electricity, however, lies the most significant legacy of the electron's discovery: it represented the first cracking of the atomic shell, considered uncrackable until then. That the electron was much smaller than the atom, that it clearly originated from within the atom (electrons were 'boiled' out of heated metal pieces, that is, out of the metal atoms), and that it was (negatively) charged while an atom was electrically neutral, established unassailable proof that atoms had to have some internal structure, and that electrons were one of their ingredients. The inference that atoms were no longer the final, indivisible and irreducible building blocks of matter, so believed until the end of the 19th century, was inescapable. This, the dawn of the atomic age as we know it today, is perhaps the most profound significance of the discovery of the electron.

The Electronic Charge

Beyond its size — or lack of it — the first and foremost characteristics of an electron are its electric charge and mass, the characteristic amount of the electric charge it carries and just how much it weighs. As the wide world of the subatomic particles came into view throughout the 20th century, the electronic charge and mass came to characterize much more than just the electron; they served to define a new standard scale for all the elementary particles that we came to know. In the next chapter, we will talk about electron mass and how it provides a new scale for mass as well as energy in terms of a new unit called the electron-volt. For now, we will focus on its electric charge.

The amount of electric charge carried by a single electron is, like everything else about it, infinitesimal by our standard, especially so when expressed in terms of units that were in place more than a century before its discovery. By the time electrons were discovered in 1897, the science of electricity had already been completed and the convention for the necessary units of measurement had all been carved in stone. Electricity, after all, is the second oldest known force of nature, second only to gravity. The scientific study of it got underway in the middle of the 18th century when in 1752 Benjamin Franklin (USA, 1706–1790) revealed the existence of two distinct types of charges — the positive and negative — and in 1785 Charles

Coulomb (France, 1736–1806) established the inverse-square law for the electric force. The electron arrived too late to claim a new unit of its own, and everything about it — its charges in particular — had to be expressed in terms of what was already there.

Most of the electrical units are household words — amperes for electric current, volts for voltage, watts for power, and so on. We have a 12-volt car battery, a 30-amp (that is, amperes) appliance, a 1,000-watt hair dryer, and so on. Perhaps the most familiar of all is the one by which we get billed from utility companies, the thing called kilowatt-hour. A watt is a unit of power, as in 100-watt light bulbs or "the ear-shattering 50-watts per channel stereo speakers!" Power is the rate at which energy is transferred from one system to another, and the product of power and the duration of the transfer gives you the total amount of energy used. The kilowatt-hour is just that, total amount of the electric energy used in the household, running everything from a 1,000-watt toaster to a cloth dryer. A watt corresponds to a power that is equal to one volt multiplied by one ampere of current. A 12-volt battery delivering 20 amps of the 'cold cranking power' puts out a power of 240 watts. Since a watt is a product of voltage and amperes, a power of 100 watts could come from 10 volts–10 amps, 100 volts–1 amp or 1 volt–100 amps source. One horsepower, by the way, is equal to 746 watts; which gives you some idea about the electric car.

An ampere, or an amp for short, is in turn a measure of electric current. Just as a unit for a river current would measure the amount of water that flows past a point in the river, the ampere measures the passage of some prescribed amount of electric charge across any given point in a wire. The standard — in fact the only — unit for charges is called a coulomb (after Charles Coulomb, who else?). More precisely, one ampere stands for the passage of one coulomb of charge in one second of time. All these units — coulombs, amperes, volts and watts — are therefore of about the same comparative magnitude. Take the 12-volt battery that puts out

20 amps of current. It moves an amount of charge equal to 20 coulombs in one second.

When the electric charge on a single electron is expressed in coulombs, it is, as expected, ridiculously minuscule. It turns out to be 1.6×10^{-19} coulombs, that is, 1.6 tenths of a quintillionth of a coulomb. Pretty darn minuscule indeed. Now, electric charges in general come in two varieties; positive and negative. Which is positive and which is negative is completely arbitrary. Benjamin Franklin called the charge (that is, the static electricity) on a rubbed glass rod positive and those on a rubbed plastic or amber rod negative. And by this convention the electron charges came out to be negative. The best known value for the electron charge is $-1.602189 \times 10^{-19}$ coulombs, that is, negative 1.602189 tenths of a quintillionth of a coulomb (see Appendix 2 for a list of the notations for the powers of ten).

The tiny amount of electric charge that resides in a single electron, all 1.602189 tenths of a quintillionth of a coulomb of it, turns out to be one of the most fundamental constants of nature. It is the irreducible unit for all transfer of electric charge among molecules, atoms and all manner of subatomic particles that we know about to date. In the case of transfer of charges among the atoms, the underlying reason for this is obvious; as the electric charge is added or subtracted from atoms, resulting in either the negative or positive ions, it is accomplished by either adding to or removing so many electrons from atoms. But even among the elementary particles not related to electrons at all, all charge transfers, from particle to particle, occur in the integral multiple of the amount of charge of a single electron; no charge transfer in any amount smaller than that has ever been observed. The minuscule amount, in so many tenths of a quintillionth of a coulomb, is in this sense a truly fundamental 'quantum' of electricity.

Whatever It Is, It Obeys a Strict Zero-Sum Rule

Over the years of studying the phenomena of electricity — some 250 years since the middle of the 18th century — we have amassed a large body of knowledge about the electric charge: the repulsive or attractive nature of the electric force between charges that obeys the inverse-square law; the electric current and how it generates heat and light; the electromagnetic principles that run motors and generators; and the rich spectrum of the electromagnetic radiation that includes among other things visible light, microwaves, and X-rays. We have mastered the principles behind all the wonders of what electric charges do. With the discovery of the electron, we have even nailed down the ultimate source of the charges. Yet, after all these years, there are a couple of questions about charges that no one has good answers for. One is the most obvious and annoying one, the kind that an innocent child might ask: What exactly is the electric charge? What is it that is characteristic of electrons that can do so much for us? The simplest answer is that no one really knows. It is one of those questions with no answer — what is space, what is time, what is mass, and so on. Many of us believe that the four — space, time, mass, and charge — are in some as yet unknown manner interrelated, but we are far from gaining even a slight hint as to how.

Another puzzling fact about the electric charge that we know to be true but have no deep insight about is the fact that the electric charge, whatever it may be, adheres to an absolutely strict zero-sum rule. Electric charges can neither be created nor destroyed, and as they are transferred from particle to particle, from atom to atom, and from molecule to molecule, the total net amount of charge before and after any reactions must always balance out. The net 'liabilities' and the net 'assets' must always sum up to zero. Known as the law of conservation of electric charge, it is one of the absolutely obeyed zero-sum rules of nature. No violation of it has ever been known. When two particles collide with each other and turn into, say, five other particles — all too common everyday occurrences in the world of subatomic physics — the sum of the charges of the initial two particles must always equal the sum of the charges of the final five particles. Suppose the two initial particles carried equal and opposite charges, the total net initial charges being zero, the sum of the charges of the five newly-created particles must always be zero. The electric charges of the final five particles may be, in the units of the electronic charge, +2, +1, −1, −1, and −1 or +3, −2, 0, 0, and −1 but not +2, +1, −1, −1 and −2. If anyone thinks he has observed the third sequence, he must have goofed up somewhere!

We can cast this conservation of electric charges in a simpler language. The fact that electric charges are always transferred in exact multiples of the electronic charge naturally leads to a simple and straightforward finger-counting technique for quantifying charges. Let's first make an analogy with money. If every monetary transaction — buying, selling or depositing — is to be done only in units of five dollars, it would be natural to expect that sooner or later the designations, five, ten, fifteen dollars and so on, would be replaced by one five, two fives, three fives and so on. The next step in the evolution would be to drop the common scale — five dollars — altogether and just refer to five, ten and fifteen dollars as one, two, and three. Everyone in the loop would know what this meant. We do the same thing with electric charges. Since the charge of a single electron is the common unit for all charge transfers, there is no point in dragging around the cumbersome number of 1.6×10^{-19} coulombs all the time. We will just invent a notation for it, say, q. All other charges can then be referred to as 1q, 2q, 3q and so on. Since the charges come in both positive and negative varieties, we would also have charges corresponding to −1q, −2q, −3q and so on. The charge of an electron in this notation is −1q. Well, you can guess the next logical step in the evolution of notation: drop the q altogether. The electric charges of particles are now encrypted by simple numbers, +3, −2, 0, +1 and so on. The original electric charge of an electron? It is −1. That simple. Now just how is a person ever to know when a simple number like +2 stands for either ten dollars or 3.2×10^{-19} coulombs? Well, you've just got to know; if you know, you know and if you don't, you don't! Just in case, however, the numbers that designate the integral multiple of the electronic charge are given a specific name. They are called the electric charge numbers (what else?).

In terms of the charge numbers — a set of simple, finger-counting numbers — the statement of the zero-sum rule of the charge conservation simplifies further: The total charge number of a system can never change in any physical processes. If the total charge number before any reaction is positive two, then no matter what happens during the reaction and no matter how many particles may be newly created as a result of that reaction, the total charge number for the final products must add up to positive two — no net increase or decrease. This is one of the most rigidly enforced zero-sum rules of nature.

The Granddaddy of All Zero-Sum Rules

The significance of this law of conservation of the electric charges became clear as we entered the age of the subnuclear world, beginning in the early 1930s and in full swing by the second half of the 20th century. As we entered this 'Wild West' of the subnuclear particles — which we will discuss later in the second half of this book — it became less and less easy for us to fully comprehend the laws of physics that govern the microcosms of elementary particles; the hordes of new and strange particles, the new subnuclear forces, and all manner of wild reactions in which particles changed their identities at will. The lesson of the electric charge and its law of conservation — that the charges can be encrypted by a set of numbers and that the total charge number cannot change — became the most important and trusted principle that guided us through the maze of subnuclear species.

Many new 'charges' were conceptualized— nucleonic 'charges,' leptonic 'charges,' and baryonic 'charges,' just to name a few—and we invoked the conservation laws for each of these new charges in exactly the same manner as for that of the electric charge. Without risking an overstatement, it can be said that most of our present-day understanding of the world of elementary particles rests on these proclaimed laws of conservation charges. These rules of the zero-sum game play an indispensable role in helping us sort out the ins and outs of the world of atomic nuclei and of the elementary particles, and the zero-sum rule for the electric charge is the granddaddy of all such rules to follow.

2

Mass: E = mc² and All That

Along with its electric charge, the other foremost thing about the electron is its mass. In listing the properties of elementary particles, the first things we specify are their charge and their mass. After all, these represent the essence of the two earliest known forces of nature — electricity and gravity. The mass of an electron, as expected, checks in with a long string of zeroes to the right of a decimal point. In terms of kilograms —the standard scientific scale for mass — it turns out to have not just a dozen but 29 zeroes following a decimal point, as in 9×10^{-31} kilograms. Rounding up, it is still some 30 orders of magnitude smaller than one. There is just no way for us to develop a sense of something that small; it is just another abstract number. In order to get a handle on it, we can try some analogies. One such analogy is this: the mass of an electron is to the mass of a mosquito what the latter is to the mass of our Sun. This helps our understanding a little, but not much.

When talking about mass, we immediately think of weight. We usually ask how much a thing weighs, not what its mass is. It is only too natural, of course; everything we feel and experience is always under the influence of the Earth's gravity. We cannot just turn gravity off! In our mind, therefore, the two are virtually synonymous; mass is weight and weight is mass. But then we also know that things do not weigh the same on the moon as on Earth, the moon's gravity being only one-sixth of that on Earth. A weight is a name for a particular force, the force of gravity tugging at an object, whereas mass is an intrinsic property of a thing.

Considering its negligibility, the mass of an electron still commands a healthy respect from scientists, and you may wonder why. Certainly, it is not because of any effects of gravity. What then is the relevance of mass, even mass that infinitesimal? The answer lies in the relationship between mass and energy that was discovered in 1905 by Albert Einstein (Germany, Switzerland, USA, 1879–1955). The equivalence relation between mass and energy, so simple and compact in its expression and so profound in its implication, is none other than $E = mc^2$, the formula that became one of the defining icons of the 20th century.

Mass and Energy

Originally introduced by Isaac Newton (England, 1642–1727) in 1687, the concept of mass was not necessarily very precise: mass was defined as a measure of inertia. But then what was inertia? Inertia referred to the stuff of matter — the quantity of matter — that tended to resist any change in its state of motion. As Newton's first law of motion states, a body in its state of rest will tend to remain at rest, and a body in its state of constant speed in a straight line (no slowing down, no picking up speed, no stopping, nor any changes in direction) will tend to maintain its state of uniform motion, unless compelled to change its state by forces acting on it. Newton himself used the terms 'mass' and 'quantity of matter' quite interchangeably: in *Principia*, published in 1687, Newton defined mass as the quantity of matter, where "the quantity of matter is the measure of the same, arising from its density and bulk conjointly." Certainly a bit circuitous, but then even Newton had to start somewhere! Fast forward to 1905 and the famous theory of relativity, and suddenly the stuffy old concept of mass was given a completely new meaning. A few decades later, we would, from this meaning, develop an understanding about the origin of the Universe itself.

When we speak of the theory of relativity, the first thing we should note is that there are two closely related but distinct branches of the theory; the so-called special theory of relativity formulated in 1905 and the general theory of relativity formulated in 1915, both by Albert Einstein. The former provides the foundation upon which is built a new theory of motion, while the latter concentrates on the force of gravity itself. The two theories of relativity closely parallel those of Newtonian physics: the first branch of Newtonian mechanics deals with the general nature of the laws of motion and the second with the force of gravity, Newton's law of universal gravitation. The general theory of relativity refers to Einstein's reformulation of universal gravitation; it involves a complicated branch of mathematics suited to describing curved space and curved time — time warps, wormholes and all that — and provides the basic tool for our continued quest to understand cosmology. The special theory of relativity, on the other hand, deals with Einstein's revolutionary new visions for mass, space and time, and it is this special theory that we usually refer to when we speak of relativity.

Almost a century after its debut, the subject matter of relativity still remains largely esoteric and perplexing. One particular formula, however, is so widely known that it needs almost no introduction. This is, of course, the famous $E = mc^2$; E for energy, m for mass, and c for the speed of light. The formula's fame is unmatched by any other in science, and spreads far beyond science's domain; it adorns everything from posters and book covers to T-shirts and stamps. It even shows up as names for corporations; one of the computer companies that specializes in storage technology is named EMC2. Then there is an investment firm on Wall Street by that name! I can just see it now — a movie with the title EMC2! Widely known, but still in need of some elaboration, especially with regard to the new meaning it gives to the concept of mass.

First off, despite its appearance — mass multiplied by the square of the speed of light — the formula in no way implies that things are flying about at the speed of light. On the contrary, one of the basic results of relativity is that the speed of an object cannot equal or exceed the speed of light. The speed of light in empty space — 300 thousand kilometers per second (or, for precision aficionados, 299,792.4562 kilometers per second, according to the latest and the most accurate measurement made by bouncing a laser light beam off a small mirror left on the surface of the moon by the Apollo astronauts) — is the ultimate speed of nature that nothing can exceed. It is one of those asymptotes of nature; with the help of some of the world's most powerful particle accelerators, the lightest known particle, the electron, can be accelerated up to a speed equal to 99.9999% of the speed of light but it can never actually reach the 'Nirvana' of the speed of light.

What then is the significance of the square of the speed of light? It is simply the proportionality factor by which mass can be converted into energy and vice versa, and this was something totally unexpected in the Newtonian scheme of things, in which mass and energy maintained their separate identities. Mass and energy, according to Einstein, were two faces of the same commodity. And it is not a small and insignificant proportionality factor either. Since the speed of light in empty space is 300 thousand kilometers per second, its square has a few zeroes to contend with; expressed in units of meters per second (that is, the square of it) it has the numerical value of 9 multiplied by 16 powers of ten. When speaking of powers of ten, it is customary to group in steps of thousands: one thousand thousands is a million, one thousand millions is a billion, one thousand billions is a trillion, one thousand trillions is a quadrillion (that is, ten to the power of 15), one thousand quadrillions is a quintillion (that is, ten to the power of 18), and so on. In this language, the square of c translates to either 90 quadrillion or 0.09 quintillion. Either way, it is a huge number, and it is by this huge factor that mass and energy convert to each other. It is a whopping exchange rate. What is so new — and singularly significant — about this formula is that it has nothing to do with whether an object is in motion or not: all it takes for a thing to possess energy — and a huge amount of it at that — is to have mass. This new form of energy is appropriately called the self energy, and this convertibility between mass and energy ranks as one of the most far-reaching of all discoveries made by Einstein: for one thing, it ushered in the atomic age, the age of nuclear power as we know it today.

Units for Energy: The Familiar and the New

The units for energy we are most familiar with are calories and BTUs (British Thermal Units): one calorie is the amount of energy (heat) necessary to raise the temperature of one gram of water by one degree Celsius, and, similarly, one BTU is the amount of energy necessary to raise the temperature of one pound of water by one degree Fahrenheit. Straightforward enough. Converting calories into BTUs, and vice versa, is another story, however. To begin with, there are two kinds of calories; the scientific one — the one just mentioned above — and the more common kind, the Calorie, that is, with a capital C. The energy value of food is specified by Calories; one Calorie is equal to 1,000 calories, that is, one Calorie equals one kilocalorie. One Calorie converts to about 4 BTUs (3.97 BTUs to be exact). Now the $E = mc^2$: if one kilogram of mass (that weighs 2.21 pounds on a kitchen table any day of the week) is allowed to be converted into energy — all of it, not just a fraction — it would turn into a formless concentrate of energy packing some 22 trillion Calories or about 87 trillion BTUs! So next time you look at a totally unremarkable sack of potatoes on a kitchen table, consider its energy potential! A total conversion of a ton of mass into energy would provide the energy needs of the United States for a full year! In actual instances, it is not such things as potatoes but rather atomic nuclei that are usually involved in the transmutation of mass into energy. A small fraction of their mass is converted into energy by a process called nuclear fission: a heavy nucleus breaks up into, say, two smaller ones, converting in the process a fraction of its original mass into energy that we call nuclear energy. In a controlled fashion, such a process generates electric power for us, but in an uncontrolled explosion ... well, you know what I'm trying to say.

The process of conversion works both ways, of course, mass into energy and energy into mass. The latter is the process by which all lumpy things in the Universe came to be what they are. Once upon a time — a long, long time ago — there was this one gigantic concentration of primordial energy that just showed up one day. In time the ball of energy just went bang in one humongous explosion. We call this explosion the Big Bang — and it must have been the biggest bang of all! After the big explosion, as the energy expanded and space began to cool down, here and there, bundles of formless energy began to 'condense' into lumpy masses, and things were created. That, at least, is how we currently understand the Universe to have come into being. A few details of the whole process still escape us, but we have very little doubt as to the origin of matter — the 'condensation' of energy into mass.

While such units for energy as Calories and BTUs serve us well in our daily lives, they are way too large for subatomic particles. We need special units for special circumstances, units for energy that are more in line with the minuscule scales involved. One such unit, relatively new, is what is called the electron-volt, eV for short. It is far from being a household word, but in the world of subatomic physics, it has become as common as a McDonald's cheeseburger. Although its name clearly invokes the electron, the idea behind it is as old as electricity itself. It came about this way.

Let's talk in terms of the more familiar notions of gravity. When an object falls to the ground from some height, it hits the pavement with the energy it gained during the fall. When a pumpkin is dropped to the sidewalk from a second-story window ... well, you get the picture. An object gains its energy from gravity by dropping from a higher to a lower level. If we borrow the term 'voltage' from electricity and apply it here, we can say that a mass (an object) has dropped through a gravitational 'voltage' (from a higher to a lower level) and gained energy during its fall. Well, the situation is exactly the same in the case of the electron-volt. The amount of energy gained by an electron as it drops through the 'height' of one volt, from a higher to a lower voltage, is defined as one electron-volt of energy. An electron pulled along a length of wire by a battery of 12 volts gains exactly 12 electron-volts of energy. Very straightforward.

Knowing how small the amount of charge on an electron is, you can guess that one electron-volt is also an infinitesimal amount of energy. It comes out to be too small, in fact, even by the standards of subatomic physics. In other words, we overdid it. We went too far down the scale, and more often than not we find it necessary to have to climb back up a little. This is why we introduced various multiples of electron-volts, such as keVs, kilo-electron-volts; MeVs, million-electron-volts; GeVs, billion (giga)-electron-volts; TeVs, trillion (tera)-electron-volts; and so on. Now, a small matter of semantics: in the cases of MeVs and TeVs, it spells the same either way — million or mega, and trillion or tera. Not so when it comes to billion and giga; one could go with either GeV or BeV. For a while both were used. Finally, the good one, GeV, won out over the bad one, BeV. So much for Beverly. Speaking of TeV, the powerful particle accelerator at the Fermi

National Accelerator Laboratory — located in a prairie about 60 miles west of Chicago — is named the *Tevatron*. It accelerates particles until they gain energy in the range of TeVs. An interesting case for notations will develop at the next levels, in the realm of quadrillion-electron-volts and quintillion-electron-volts. A simple QeV will not do; QdeV and QteV? How are the electron-volts related to the more familiar Calories? We would have to throw in a long string of zeroes. One Calorie, that is, one thousand calories, corresponds to about 26×10^{21} electron-volts. That is, 26,000 quintillion electron-volts. Gives some vague idea about just how tiny an amount one electron-volt of energy is, doesn't it?

How Much Does a MeV Weigh?!

The popularity and usefulness of the electron-volt comes not necessarily from its small scale but rather in a way that is not immediately apparent: it provides a very convenient way of defining units for both mass and energy, banking on the formula $E = mc^2$. The relationship $E = mc^2$ can be turned the other way around to spell a definition of mass in terms of energy, as in $m = E/c^2$. Sounds so trivial, but this is the reverse of the usual; in most cases, the energy of an object is a derived quantity that is predicated upon its mass; mass first and energy in terms of it. The 'new' discovery, $m = E/c^2$, gives us a handle by which we can do the reverse, take a value of energy and then define mass in terms of it. There is one catch; in order for this to work, we have to have a way of quantifying energy without having to define mass first. Well, the electron-volt is a perfect fit. It is a unit for energy that depends only on things that are electric; the charge of an electron and voltage. Combined with $E = mc^2$, it provides for a system of units in which conversion of mass into energy and vice versa is as smooth as silk.

It is really all very simple; we express the mass of a particle in terms of its self energy mc^2. The simplicity of this can catch you off guard.

Suppose we have a particle whose self energy mc^2, is, say, 10 million electron-volts, 10 MeV, we express its mass as — are you ready for this? — 10 MeV/c^2. The numbers stay the same; it is the units that change, from MeV to MeV/c^2, from electron-volts to electron-volts divided by c^2. In this language, mass is just so many eV/c^2 and its self energy is the same number in eV.

Now, we are in a position to list the properties of an electron in the cryptic format that is the standard of the trade:

Name	Electron
Symbol	e
Charge	–1
Mass	0.511 MeV/c^2
Discovered	1897

The charge of –1, of course, stands for –1 of q, where q is the amount of electric charge on a single electron, that is, 1.6×10^{-19} coulombs. The mass, rather than being listed as 9×10^{-31} kilograms, gets a new facelift — about a half million-electron-volts divided by the square of the speed of light. Somehow 0.5 sounds more user-friendly than a number with a string of 30 zeroes after a decimal point, and, likewise, –1 is much more palatable than 10^{-19} coulombs. It is all in the expressions!

3

The Photon: No Charge, No Mass

Three years after the discovery of the electron, in 1900, another epoch-making discovery came to pass. Five years later, in 1905, the new discovery was christened with a new name, the photon, meaning the particle of light. A casual perusal of a dictionary reveals this gem of a definition: a photon is "a quantum of electromagnetic radiation, usually considered as an elementary particle that is its own antiparticle and that has zero mass, zero charge, and a spin of one." Whoa! Not so fast! Not in one breath! This is quite a mouthful even for seasoned pros, and almost completely opaque to the uninitiated. It throws out at least a half dozen strange words — quantum, antiparticle, zero mass, zero charge, and the mysteriously cryptic "a spin of one." One what? It will take some doing before one can begin to make sense of this definition. Perhaps we should start by talking about the most familiar item in that statement; good old electromagnetic radiation.

Electromagnetic Radiation

This is old hat. Everyone knows what it is. From its lowest low end to its highest high end, electromagnetic radiation covers a wide spectrum. In much the same way that a long stretch of a boulevard is known by several different names along its route — each claiming a portion of it several blocks long — electromagnetic radiation is known by many 'local' names. Starting at the low end, we have the AM radio band, the amateur ham radio band, the VHF TV band, the FM radio band, the UHF TV band, followed by the short-wave radio band and the radar band. The midsection covers microwaves, infrared, visible light and ultraviolet rays. Soft and hard X-rays and gamma rays top off the high end.

Each band claims a few 'blocks' of frequencies. The frequency — the number of cycles per second — is old hat and is expressed in units of Hertz; one Hertz (Hz) being one cycle per second. The standard US household currents come in 60 Hz. The spread of the electromagnetic spectrum covers some 20 orders of magnitudes in frequency, from a mere 100 Hz for the long-wave region (the walkie-talkie wave) to ten billion terahertz (10^{21} Hz) for some of the most powerful gamma rays observed in space. The product of frequency and wavelength is, by definition, equal to the speed of a wave; if your stride (wavelength) is 3 feet long and your pace (frequency) is 5 strides in one second, you are moving at the speed of 15 feet per second. Because electromagnetic radiation has, at any frequency, only one unique value for its speed (the speed of light), the frequency and wavelength of any given band are inversely correlated; the lower the frequency, the longer the wavelength and, conversely, the shorter the wavelength, the higher the frequency.

Electromagnetic radiation has energy and momentum of its own. The energy of radiation is obvious; its omnipresence is virtually its own definition — solar energy, microwave ovens or just the plain warmth of sunlight, to name just a few. On the other hand, the notion of radiation momentum may sound a bit remote. After all, no one ever heard of someone getting knocked down to the ground by a beam of light, no matter how bright! Even for those blinding stadium floodlights, the idea of light having its own momentum is definitely less than convincing. It is so only because, in our scale, radiation momenta are too minuscule to matter. For subatomic species, such as electrons, however, the momentum of light is no laughing matter at all; radiation momentum will knock atomic electrons off their tracks any time of the day — we call this ionization. To drive home the importance of radiation momenta, some scientists went so far as to toy with the idea of space travel based on radiation momentum as a rocket propellant! Since the thrust of a rocket comes as a reaction to the action of the ejected momenta of burning fuel, why not a rocket whose thrust comes from the ejected momenta of beams of powerful light? Just turn on the tail lights and whoosh! That at least is the idea behind so-called 'photon rockets.' There is nothing unsound about the principle — it's just not very practical.

Now, to be sure, there are some subtle differences between the energy and momenta of radiation, on the one hand, and those of ordinary matter, on the other. For one thing, radiation has neither mass nor electric charge, the two basic characteristics of particles such as the electron. For another, a particle is a pinpoint concept, localized at a point at a given moment of time, whereas radiation is just the opposite, spread out all over in all directions. Those, at least, were some of the classic distinctions between matter and radiation that persisted for more than three centuries. Suddenly, a discovery was made in 1900 that brought this clean separation between the two to a screeching halt. As the result of that discovery, matter and radiation became a little less distinguishable from each other, each picking up traits of the other, and things were never the same again.

Planck's Quantum

In the several years prior to the year 1900, many scientists studied the nature of radiation energy in great detail; in particular, how smoothly continuous its energy variation was vis-à-vis other characteristics of radiation such as amplitude, frequency, wavelengths, and so on. In 1900, Max Planck (Germany, 1858–1947) came upon a result that was hard to believe at first: radiation energy — the warmth of sunshine — was not creamy-smooth. It turned out to be lumpy. Upon closer examination, radiation revealed its hidden secret: its energy was calibrated in terms of its own indivisible scale — the smallest quantity of radiation energy, which could not be further reduced. An often-invoked metaphor is that of a beach and its sand. Picture a smooth and flat expanse of a beach, stretching smoothly into a horizon, with dunes here and there. Of course, we know that there is nothing really continuous and smooth about it; the beach is covered by tons upon tons of sand, zillions upon zillions of sand grains. Each grain of sand is irreducible. What Max Planck discovered was that radiation energy was made up of zillions upon zillions of energy 'grains' — tiny, separate and individual 'grains' of energy.

Planck himself didn't quite know what to make of it at first, but that did not prevent him from promptly naming it. Planck called it the quantum, that is, the quantum of electromagnetic radiation. The word would forever change the face of science in the 20th century; it describes everything from quantum mechanics to the quantum technology of today — quantum physics, quantum chemistry, quantum electrodynamics, quantum field theory, quantum electronics, quantum semiconductors and so on. According to this new 'quantum' view, the energy of radiation is determined by two things — the total number of quanta and the energy content of the quantum. Let's compare a quantum to a copper penny. The total amount of copper (the energy) in a sackful of pennies depends on two things: the number of pennies in the sack (the number of quanta) and the amount of copper that is in a single penny (the energy content of a quantum). If a penny is made with one gram of copper, a sack containing 1,000 pennies corresponds to one kilogram of copper. If, on the other hand, a penny of some other currency is made with two grams of copper, a sack of 800 of these pennies corresponds to 1,600 grams of copper.

The discovery that radiation energy was granular, discrete and countable was in itself truly revolutionary at the time, but there was more to it. Planck successfully discovered the quantitative relation that determines the energy content (the copper amount) of each quantum (the pennies). It was something totally new: the energy content of a quantum was related, of all things, to the frequency itself. The higher the frequency of radiation, the greater the energy of its quantum. Some pennies are larger or smaller than others, the larger ones (of higher frequency) containing more copper. The energy (intensity or brightness) of radiation depends on two things; how many quanta is it made of and what its frequency is, since it is the frequency that determines the energy value of each quantum.

Now, to be sure, the quantum of light corresponds to an unspeakably infinitesimal amount of energy. The energy flux of sunlight reaching the surface of the Earth on an average sunny day is about 1,000 watts per square meter (you can now extrapolate from this to arrive at a rough guesstimate of just how much electric power it takes to put on a night game) and this works out to about two and a half billion trillion quanta hitting an area of one square meter on the ground every second. The individual granularity of the warmth of sunlight is just too fine a detail for us to sense directly. How much energy does a single quantum of light actually pack? Let's pick a blue-green light with a wavelength at about the midpoint of the visible light spectrum, say, 500 nanometers. For this wavelength, the energy of a single quantum comes out to about 2.5 electron-volts. Expressed in electron-volts, one can see that the energy of light quanta is in the same ballpark as the energy scale of things in the subatomic world.

Einstein's Photon

The stage now moves forward five years to 1905, the year of relativity. We have already met the energy-mass equivalence formula, $E = mc^2$: it relates mass to energy and conversely energy to mass. It applies to all things in the Universe that have mass. There are no exceptions to it. But then, what about the possibility that 'things' could very well exist without any mass? Within the framework of pre-Einsteinian physics, this question was for all intents and purposes a non-question; things are things because of their substances, and mass is that substance. No mass, no substance, and no thing! Can one talk about things if they do not have mass to be defined by? This question-statement was answered in the Einsteinian theory of relativity in the affirmative. Yes, it is true that the formula applies only to objects that have mass; mass can be infinitesimally minuscule but as long as it is not exactly zero, the formula applies. By the same token, a particle with mass, however small, cannot be accelerated to a speed equal to the speed of light. Things can attain speeds that get ever closer to the speed of light — 99.999998% of it — but cannot reach the ultimate speed. Having mass and reaching the speed of light are mutually exclusive and absolutely forbidden.

What about the flip side of it? Is it possible for something to exist that has no mass at all? Relativity answers this with a resounding yes, with one unalterable condition: yes, it is perfectly possible for things to exist — by that it is meant first and foremost that they carry energy and momentum — without any mass at all, provided they meet one stringent requirement: they exist in a perpetually moving mode, moving only at one speed; the speed of light. Relativity presents these two camps that are mutually exclusive: no mass moving only at the speed of light, on the one hand, and any non-zero mass unable to attain the speed of light on the other. In the latter camp belongs just about everything that we know about, but we must bear in mind that relativity does make room for the other reality — zero-mass moving at the speed of light.

Having provided for such an extreme reality, however, relativity now backs away from the question of what energy and momentum should be for the zero-mass particles. The theory, for all its insight, cannot tell us what the energy and momentum are for such particles. That information will have to come from sources other than relativity. It tells you that the idea of the zero-mass particle is a perfectly tenable one, but does not give you the quantitative definitions for it. It is not exactly an everyday concept — a particle without mass that moves at the speed of light! A marble without glass that zips across space at 186,000 miles per second!

Once Einstein realized this possibility, it didn't take him long to see the connection between it and what Planck had discovered five years earlier. Electromagnetic radiation, which by definition moves at the speed of light, carries its own energy and momentum, and as Planck showed, it was to be viewed as streams upon streams of discrete, indivisible units of energy that Planck named quanta. Immediately, everything clicked. To Einstein the quantum of Planck was not just an energy unit of light; no, it was much more than that. It was a perfect example of the massless particle: it was a particle because it had a basic unit of energy, and momentum, but it had no mass and always moved at the speed of light. After all, radiation does not have any mass and is, by definition, moving at the speed of light. Upon this realization, Einstein promptly renamed Planck's quanta 'photons,' and elevated them to the ranks of full-fledged particles: if the electron was the particle of electricity, then the photon was the particle of light. No mass and no charge, but a particle nevertheless. The metamorphosis of Planck's quanta to Einstein's photons was more than a mere change in name; it represented a major shift in our thinking about the definition of a particle, from what it is (mass, charge, location, speed, etc.) to what it does (energy and momentum). With the vision of photons as massless particles, Einstein did away once and for all with the concept of a particle as dependant upon its having mass: a particle was now a particle as long as it did like a particle, whether it had mass or not.

The list of the basic properties for the photon, in the same cryptic notation, looks like this:

Name	Photon
Symbol	γ (gamma)
Charge	0
Mass	0
Discovered	1900/1905

The idea of a particle with no mass always runs into a ton of psychological resistance when first mentioned. A particle without mass sounds as crazy and empty as a marble without glass — so counter-intuitive. After all, our intuition is acquired through our daily experiences and observations and it goes something like this: first, there are things, things by definition have mass — things weigh, don't they? And while something may be electrically charged most are naturally electrically neutral ('zero charge'). The whole idea of a particle with zero mass just doesn't go down smoothly. Actually, you are in for more: photons were not the last of the 'vanishing mass' kind. Down the road, we will meet at least three more subatomic particles that have no mass — three different kinds of particles called neutrinos.

4

The Spin: If It's Round, It Rolls

In any listing of elementary particles, the third item to be specified, after electric charge and mass, is something called 'spin.' The name conjures up an image of something completely familiar — a spinning top, a fast pitch of a spinning baseball, or, for that matter, the rotation of the Earth about its own north-south axis. Such metaphors are routinely used to describe spin, for a very good and desperate reason: as we will now discuss, the thing we call spin has no known counterpart in our human-sized world. Charge has electricity and mass gravity, but there is nothing in our scale that resembles even remotely what the spin of a particle is really like. As it turns out, throughout the development of the science of elementary particles, we have encountered many such quantities; physical attributes that are defined only within the reduced dimensions of the microscopic subatomic world. Spin is just the first of many such hard-to-understand attributes, and in order for us to describe it we have little choice but to fall back on convenient metaphors, such as the rotation of the Earth.

The idea of the spin, or more specifically the spin-like behavior of electrons, was first proposed in 1925. The two decades between 1905 — the year of relativity and photons — and 1925 — the year of quantum mechanics and the spin — were unmatched by any other such periods in the history of science. It was a time of unparalleled achievements, one monumental discovery following another in breathless succession — the planetary structure of atoms, the discovery of atomic nuclei, and the birth of a new physics called quantum mechanics. And the idea of spin turned out to be one of the most critical components of the atomic theory that provided the basis for understanding the pattern of recurring similarities among the elements of the periodic table.

A Picture of the Atom Emerges

If the first decade of 'the 30 years that changed the world' was marked by the discovery of the electron, the quantum/photon and the theory of relativity, the second decade — the 1910s — was the time of the discovery of the secrets of the atomic structure. Two giant achievements during this period stand above all others — the discovery in 1911 by Ernest Rutherford (New Zealand, Canada, England, 1871–1937) of the atomic nucleus, and the first successful (albeit semi-quantitative) theoretical model for the structure and workings of the hydrogen atom in 1913 by Niels Bohr (Denmark, 1885–1962). The discovery by Rutherford established once and for all the so-called planetary model of the atom: an atom consists of a positively-charged central core (the atomic nucleus) and a group of electrons orbiting around it, in much the same way the planets orbit about the Sun.

Building on this planetary model of atoms, the first working model for the simplest atom — the hydrogen atom — was put together in 1913 by Bohr. This so-called Bohr model of the atom was far from being a full-fledged theory, but was nevertheless a semi-quantitative framework in which some calculations could be performed and their results compared to actual experimental data. In a deft synthesis of the ideas and discoveries of Thomson, Planck, Einstein and Rutherford, Bohr managed to put it all together —

electrons, photons and the atomic nucleus — into one unifying working hypothesis. At the dead center of an atom sits its nucleus pulling on all orbiting electrons. The orbits are separate and distinct, expanding out in concentric circles from the innermost one to the outer edge of the atom.

Normally, the electron occupies the innermost orbit — the lowest rung of the atomic ladder — and does not venture out to explore the higher 'rungs.' When an unsuspecting photon skirts by the electron (when light is shone on an atom), the electron wakes up and will gulp down the photon and with the extra energy gained from the photon it will jump to a higher orbit — a higher rung of the ladder. This was the 'new' atomic view of the process of absorption of light by matter. The other side of the coin — the emission of light by matter — was just the reverse: sooner or later, the electron now perched up on a higher rung will jump back down to its natural habitat, the lowest rung, this time coughing back out the extra energy in the form of a photon (emission of light by an atom). Things get a little more interesting when the energy of the absorbed photon is large enough for an electron to make a jump, from the first to, say, the fourth rung, in one leap. It can jump back down to the lowest rung in one leap, the same way it went up, or take its own sweet time in coming down in one of three ways: first down to the third rung and then making a smaller leap down to the first; leaping to the second and stepping down to the first, or coming down one rung at a time in three steps. Each mode will cough out two or three photons of different energies.

What now appears to be a simple picture that makes all the sense in the world was a completely revolutionary proposal when Bohr made it. But the way the model yielded numbers that agreed with the observed spectra — the absorption and emission light spectra of the hydrogen atom — was unassailable and established the Bohr model as the guiding light for peering into the structure of all other atoms.

The Secret of the Periodic Table

Emboldened by the success of the Bohr model, the development of the new physics of quantum mechanics went into overdrive in the 1920s. A crude model gave rise to searches for underlying principles and a framework for a mathematical theory built on those principles. As the ideas of quantum mechanics compacted into a well-established theory, and the knowledge of atomic structure expanded from the simple hydrogen atom to the family of heavier many-electron atoms, the explanation of the recurrence patterns collectively known as the periodic table of the elements became the acid test of the new physics. In the periodic table, the elements are grouped into nine distinct groups by their chemical similarities. In one group, hydrogen, lithium, sodium, potassium and others share common chemical behaviors; such inert gases as helium, neon, argon and xenon form another group; carbon, silicon and germanium belong to yet another group, and so on.

In 1925, searching for the key for the recurring regularities among the elements, a young 25-year-old by the name of Wolfgang Pauli (Austria, Switzerland, 1900–1958) made a brilliant deduction: each potential orbit for the electrons within an atom (the rungs of the ladder) can be maximally occupied by no more than two electrons. An orbit may be unoccupied and empty, it may house only one electron or be completely filled by two electrons, but under no circumstances will it have more than two. The next electron, if an atom should already contain two electrons, would have to move up and occupy another orbit, the next rung up the ladder. According to this inspired principle — the so-called exclusion principle — the chemical similarity between hydrogen and lithium comes about this way. The hydrogen atom has just one electron that occupies its natural orbit (the lowest one). The lithium atom, on the other hand, contains three electrons; the first two having completely filled out the lowest orbit, the third one must necessarily be housed in the next higher orbit. Looking in from the outside, the first thing other atoms see in both cases is thus the lone outermost electron, the only one in the case of hydrogen and the third 'lonesome end' in the case of lithium. This is the reason why the two behave similarly in their chemical dealings with other atoms.

Now, in order for Pauli's exclusion principle to make sense, a couple of hitherto unknown aspects of electrons had to be assumed. Firstly, the electrons possessed some new basic attribute, beyond their charge and mass, that had only two possible values — plus and minus, yes and no, long and short, or on and off. Secondly, with respect to this new 'binary' property, the electrons were 'exclusive' of each other in that they did not allow themselves to be completely identical to each other, at least within the confines of one and the same atom. The two hypotheses combined provided a natural explanation for the exclusivity of electrons: one cannot place a third electron in an orbit which is already occupied by two electrons, because the third one would have to be identical to either of the two. Let's say the two-valued property is labeled by 'long' and 'short.' The third electron, be it 'long' or 'short,' would be identical to either the 'long' one or the 'short' one already there in the orbit. When Pauli proclaimed the now-famous exclusion principle, the notion of any two-valued attribute for the electron was not yet hatched; the principle was an inspired guess deduced from the study of the periodic table. The idea of just such a two-valued attribute followed just ten months later.

The Long and Short of the Spinning Dot

Ten months after the proposal by Pauli — in November of 1925 — the other shoe dropped. Two young Dutch physicists, George Uhlenbeck (Netherlands, USA, 1900–1989) and Sam Goudsmit (Netherlands, USA, 1902–1978) made what turned out to be a major and critical discovery in quantum physics: the electron did indeed possess a totally new 'binary' attribute — it had a 'spin,' as if it were rotating about its own axis. More importantly, this 'spin' had only two ways of 'rotation,' either clockwise or counterclockwise. That was the binary code by which one electron distinguished itself from another, one turning one way and another the other way.

The simplest — and, in fact, irresistible — metaphor for spin is the rotation of the Earth about its axis, the night-and-day rotation. As it orbits the Sun, the Earth turns on itself. So does the moon; it spins as it orbits the Earth. Every known star and planet spins, as if it is their birthright. It is a striking parallel — why not an electron also? There is a bit of a caveat here, however. Since the electron has virtually no size to speak of, the phrase 'spinning about its own axis' is to be taken with a pinch of salt. Rather, the aspect of the electron that comes closest to the description 'spin' derives from its magnet-like behavior when bathed in a strong magnetic force field: an electron behaves like a tiny bar magnet and in this magnetic environment it exercises its two-valued option — with its 'north' pole pointing in one way or in the opposite direction. We speak of the 'spin' being 'up,' rotating clockwise and the 'north' pole pointing 'up,' and the 'spin' being 'down,' rotating counterclockwise and the 'north' pole pointing 'down.' In terms of the up-or-down option of the spin, the exclusion principle of Pauli gets rephrased thus: everything else being the same (in one and the same orbit), two electrons manage to keep their exclusivity by one being spin up and the other spin down, never being totally identical clones of each other.

Even if it had turned out that the spin was an attribute exclusive to electrons — no other particle owned up to it — and that the only function it served in the scheme of things was to provide the basis for the exclusion principle to help arrange all known elements according to the periodic table, it would still have been more than enough for spin to claim its rightful place among the most fundamental of all physical properties. After all, it underwrites the periodic table of the elements and, by extension, the structure of all matter in the Universe. But, as it turned out, electrons weren't the only ones that had spin. Photons — oh yes, the no-mass-no-charge photons — also turned out to have spin, all their own.

A Photon Does What?!

Yes, indeed, photons also turned out to have spin. Now, it is a little hard to visualize something that has neither mass nor charge to be associated with any kind of 'spinning' motion, but it is not really as strange as it might first seem. The spin of the photon is deduced by observing the spins of electrons, before and after their having come into contact with photons. When atoms emit or absorb light, they do it just the way we described above — the electrons emit or absorb photons as they jump up and down the various atomic orbits (the rungs of the ladder). By determining the spin of the electron before and after such jumps, we can conclude that photons either take away from or bring to the electron fixed amounts of spin. It is an indirect inference but just as real nevertheless. In a game of billiards, suppose you feel an invisible gust of wind that blows across the billiard table and strikes a ball head-on. Let us suppose further that the struck ball suddenly started spinning wildly, you would say that the gust of wind had a spin! It is like that; electrons and photons constantly exchange their spins this way.

There is one marked difference, however, between the spin of a photon and that of an electron, and it is in their respective amounts: photons carry spin exactly double that of electrons. What is meant by the amount of spin is like the case with the amounts of electric charge; designating the amount of the electric charge of an electron by the letter q, all others are expressed in this scale, as

2q, 3q, and so on. Similarly, if we designate the amount of the electron spin by a letter s — how fast it 'rotates' — the spins for other particles can be expressed as 2s, 3s, and so on. In this notation, the spin of a photon is determined to come in the amount of 2s. The fact that the photon spin comes out to be twice that of an electron is simple to understand. Let's say that the clockwise spin of an electron — the spin 'up' — corresponds to a positive one and the counterclockwise spin — the spin 'down' — to a negative one. No zeroes are allowed since that would mean an electron that has no spin, and there is no such thing; an electron always spins one way or the other. A photon that is being emitted or absorbed by an electron, as far as the spin of the electron is concerned, can do one of two things — either leave it alone or change it, either from +1 to −1, or from −1 to +1. Photons are observed to be always changing the spin of the electron that they are emitted from or absorbed to, and either way they must subtract or add by 2 units — subtract 2 to change from +1 to −1 or add 2 to change from −1 to +1.

In the nomenclature of the units for spin, a historical switch was defined in a reverse fashion: instead of 1s for the electron spin and 2s for the photon spin, the spin of the photon was designated as being one and consequently the spin of the electron as being one-half of that of the photon spin. This became the standard, and thenceforth the electron spin is always referred to as the half-integer spin and the photon spin the integer spin. Alternatively, we could have called the electron spin the odd-integer spin and the photon spin the even-integer spin. We will follow the normal usage and take the spin of a photon to be the standard. Let's recall a definition of the photon quoted at the beginning of the previous chapter; a photon "… has zero mass, zero charge, and a spin of one." You see, it took a whole chapter to shed some light on the last four words — "a spin of one."

The list of properties for the electron and photon now picks up a new entry, the spin, and in the usual cryptic manner it looks like this:

Name	Electron	Photon
Symbol	e	γ (gamma)
Charge	−1	0
Mass	0.511 MeV/c^2	0
Discovered	1897	1900/1905
Spin	1/2	1

The three quantities — charge, mass and spin — are the most basic of all properties attributed to a particle. As we mentioned in the beginning of this chapter, the spin remains a concept that is far removed from our familiarity, for the simple reason that unlike mass and charge the quantity called spin does not add up to magnitudes that we can recognize in our daily world.

5

Antimatter: A Mirror Image

Antiparticle, antimatter, antiuniverse ... This is the stuff — along with such things as time travel, space warps, and 'the other side of a black hole' — that nourish imaginations on the cutting edge of science fiction. Antimatter, however, is something very real. True, we don't run into it often on the street — we don't exactly find a shelf of antibread at a neighborhood grocery store — but for the elementary particle specialist antimatter is as real as his morning cup of coffee. The gigantic machines called high-energy particle accelerators synthetically produce antimatter in such abundance that more often than not they are a nuisance rather than a novelty.

Our knowledge of antimatter goes back 70 years; the need for such a 'crazy' idea was theoretically proposed in 1928 and the reality of it confirmed experimentally in 1932. One might say that antimatter was the last great discovery that rounded off 'the 30 years that changed the world,' the first three decades of the 20th century. Combined with the energy-mass equivalence formula, $E = mc^2$ — discovered 25 years earlier — the reality of anti-matter helped to lay the foundations of our current understanding of the very origin of matter and, by extension, the origin of the Universe itself.

An Anticlone of an Electron

By the year 1927, the new physics of atoms called quantum mechanics was virtually completed. It was all too natural for scientists to attempt to find a common framework in which both relativity and quantum mechanics could be cast in a single form; quantum physics cast in the new conceptual framework of space and time as constructed by the theory of relativity. The search for such a unified format, called relativistic quantum mechanics (certainly rather a mouthful), began in earnest. Many attempted it, but all proposals suffered from one undesirable side effect or other; one such side effect, for example, required the introduction of the notion of a negative probability, a probability for chance that is less likely than no chance! No can do!

One form of the relativistic quantum theory, put forth in 1928 by Paul Dirac (England, 1902–1984), stood out among others in that it was a simple and compact way to bridge the two disciplines in an elegant manner, without any preposterously undesirable side effects. The new proposal, however, required one critical prerequisite for it to stand up; it would need a kind of particle that was totally new. The so-called relativistic quantum theory of Dirac would be complete only if we allowed for the existence of a particle that was the 'mirror image' of an electron, an exact clone of an electron in every aspect save one. Its mass and spin would be identical to those of an electron, but its electric charge would be the exact opposite, that is, of exactly the same amount but positively charged. An interesting idea. There was just one problem; at the time Dirac proposed his idea, there was nothing that even remotely resembled such an entity. Now, you might ask: what would such a 'positively-charged' electron look like? The way oppositely-charged particles behave as mirror images of each other shows up best when they are bathed in a strong magnetic force. The magnetic force provides a strong crosswind either from right to left or from left to right; an electrically-charged particle flying through a magnetic force field would curve either to the right or left depending on the sign of its electric charge, and a pair of identical but oppositely-charged particles would behave exactly like the mirror images of each other, spreading out in both directions, just as a pair of identical cars would go separate ways at a fork in the road. Well, in 1928, no such 'spreading pairs' were to be found anywhere.

Four uncertain years passed until the dramatic vindication of the idea came one day in 1932. In those early years, much of the data for particles were collected by sending up photographic plates in high-altitude balloons. Interstellar space is filled with extremely high-speed particles that zip across the void at all times. They are named cosmic rays — beams of particles of cosmic origin — and when they collide with atoms of gases at the edge of the outer atmosphere, they recreate, out of their violent collisions, all sorts of other particles, both old and new. Before the advent of the high-energy particle accelerators that recreated these reactions artificially in controlled settings, cosmic rays' reactions were the only source of studying the goings-on in the world of subatomic particles — by catching their tracks on the plates as they rained down toward the surface of the Earth. It was in one of these photographic plates that Carl Anderson (USA, 1905-1991) discovered in 1932 the unmistakable telltale signs of curving pairs — pairs of electrons and their matching positive clones curving and curling in perfect mirror-image patterns.

Named the positron, the positively-charged clone of the electron was exactly as Dirac had predicted. It had the same mass and spin as the electron; it was a 'carbon copy' of an electron except for one thing — its electric charge was the exact opposite; it was the 'anticlone' of the electron. Often we run into a mistaken notion that the positron has a negative mass, as opposed to the electron's positive mass. Not true. There is no such thing as negative mass in the Universe; what is 'anti' about a positron is that its electric charge, not its mass, is opposite to that of the electron. The pair look like this:

Name	Electron	Positron
Symbol	e^-	e^+
Charge	−1	+1
Mass	0.511 MeV/c^2	0.511 MeV/c^2
Discovered	1897	1932
Spin	1/2	1/2

Antimatter, the Big Bang and All That

The fact that a clone of a particle has an electric charge opposite to that of the particle itself may not warrant the rather confrontational designation 'anti.' When we have a pair of twins — one boy and one girl — no one would call one the 'anti' of the other. Here we have two twins — an electron and a positron — so it should take more than just having opposite charges to call them the 'anti' of each other. In fact, the 'anti-ness' of the positron lies not so much in its positive charge nor its being the mirror image of an electron in the presence of a magnetic force field, but rather in the dramatic and violent way a positron and an electron greet each other.

When a positron and an electron come together, the most extreme act of self-destruction occurs; they annihilate each other totally and completely and both disappear in a 'puff of smoke,' that is, in a flash of light. Since one single photon cannot balance out the incoming momenta of the electron and the positron, the annihilation process produces, actually, two flashes of light — two photons shooting off in opposite directions away from the point of the total conversion of substance into radiation. In this act of 'kamikaze' bravado, the opposite electric charges cancel each other out and, more importantly, the masses of the pair turn into energy exactly according to the formula $E = mc^2$. A matter and its antimatter come together, and in blinding flashes of light, turn themselves into formless energy. It is this act of total annihilation that defines matter and antimatter to be the mirror images of each other, the 'mirror' being the radiation energy, namely, the photons.

Now the process of annihilation has its inverse. Under the right conditions, the radiation energy — a photon — will convert itself into a particle and an antiparticle. Since photons are electrically neutral, the electric charges of the pair come out exactly opposite to each other; a zero breaks up into +1 and −1. The masses of the pair are created by the 'condensation' of energy in accordance with, again, the relation $E = mc^2$. This process, called the pair creation, occurs all the time in great abundance inside high-energy particle accelerators, and is, in fact, one of the most widely-utilized techniques of artificially producing antimatter.

For both of these processes — annihilation and creation — the photon is the neutral reference point, that is, the matter-antimatter neutral; it is the mirror with respect to which matter and antimatter are mirror images of each other. It is in this sense that the assertion, "a photon is its own antiparticle" — quoted at the beginning of a previous chapter — while not being inaccurate, is something of an overstatement. A more accurate way of saying this is that a photon is the reference line that divides matter and antimatter, and that the distinction between a particle and an antiparticle does not apply to photons; there is no such thing as antilight, which when combined with light produces darkness!

The dance of creation and annihilation — energy into matter and antimatter, and their annihilation back into energy — is a powerful confirmation not only of the existence of antimatter but in fact the fundamentality of the relation $E = mc^2$ that underscores it. And this process of creation and annihilation goes on endlessly. Not only does it go on all the time in laboratories, but in fact this is the basis of our latest understanding of the origin of the Universe itself: Once upon a time came the beginning of time, and out of nowhere a point of infinite energy appeared, and in no time, it just went 'bang' in one gigantic explosion. In the thick soup of an infinitely hot and infinitely dense concentration of energy, all three — energy, matter and antimatter — coexisted, transforming themselves back and forth between radiation and substance. As the primordial point — the point of origin of the Big Bang — began to expand out, things began to cool off — slightly cooler than being infinitely hot — and along the expansion, here and there, bundles of energy 'condensed' into substances; the newly-born substances aggregated to form larger and heavier substances that attracted more substances to them, and, well, the rest is history. That, at least, is our current understanding of how the Universe got started.

There is one caveat here, however. When energy turns into substance, it is done by the process of pair creation — a photon turning into one part matter and one part antimatter. If this balance was followed strictly, we should have had a perfect balance of matter and antimatter, from the first primordial Big Bang down to the present-day Universe, and at all times in between — equal parts of matter and antimatter. Well, things did not exactly turn out that way, did they? Matter is all around us — including our very own selves — and antimatter is virtually non-existent in any natural form anywhere in the Universe as far as we can tell. How did this imbalance come about? No one knows for sure, but that never stopped anyone from coming up with a good 'explanation.' The way we 'explain' it goes something like this: as the fluctuating dance of pair creation and annihilation went on in the early moments of the creation of the Universe, somewhere along the line, for some as yet unknown reasons, an initial imbalance, however slight, crept in. Rather than a perfect 50-50, the production of matter was favored by a small fraction over that of antimatter, say, 50.001% over 49.999%. As the chain of creation, annihilation, recreation and re-annihilation continued, the imbalance, negligible at first, became larger and larger and, well, finally the Universe as we know it today became all matter. This is our best shot at this uncomfortable question. Why the initial imbalance? Divine intervention? No one really knows.

Because of the overwhelming preponderance of matter over antimatter in the natural world, it is extremely difficult to hold onto the artificially-produced antiparticles for any meaningful period of time. As soon as we create pairs of particles and antiparticles, the latter are immediately annihilated by matter particles in the walls of whatever container or chamber they are produced in. As the antiparticles disappear into the walls, all we are left with are hundreds upon hundreds of flashes of light. Antimatter, while produced with relative ease, is extremely difficult to hold onto, so we might just have to postpone indefinitely any hope of seeing a loaf of antibread at our neighborhood grocery store.

6

The Nucleus: A Whole New Ball Game

The early 1930s represented a transition period between two eras in the history of physics; one was drawing to its close and another was just getting underway. 'The 30 years that changed the world' that included the discoveries of the electron in 1897, the photon in 1900, and the atomic nucleus in 1911, were coming to an end. Profound accomplishments unmatched in any other such periods in the history of physics— the new knowledge and insights gained from the theory of relativity on the one hand, and the new principles and mathematical tools in the form of quantum mechanics on the other — enabled us to understand, as never before, the workings of the subatomic world. We now knew why atoms are the way they are, and how atoms and molecules interlock with each other to form bulk matter in the Universe. The confirmation of antimatter— the discovery of the positron in 1932— was the 'icing on the cake' of the glorious '30 years' that produced atomic physics as we know it today.

It was also the time for a beginning; the beginning of new revelations from deep inside the confines of atomic nuclei, and our realization that within the walls of nuclei lay hidden a whole new world of physics — new playing fields, new players and new rules. It was a whole new ball game. It was the beginning of nuclear physics — the science of nuclear structure, nuclear reactions and nuclear power — and, later, elementary particle physics. It would be an entirely new journey. Having completed the study of the 'planets,' we were now poised to take a peek into the innards of the 'atomic sun,' the nucleus.

Further Than Pluto's Moon

To be sure, the planetary picture of an atom with the nucleus at its dead center was something that had been known since 1911, fully two decades before the beginning of nuclear physics per se in 1932. All throughout the development of atomic physics, however, the atomic nucleus played the role of a passive dot at the center of an atom. It had more than 99.9% of the mass of an atom, positive electric charges in amounts equal to the sum of the negative charges of atomic electrons, and was an 'immovably' heavy central anchor around which electrons gathered and whirled around in circles. As far as the physics of atoms was concerned, it was just a point that served to define the center of atoms; the details of the 'points' themselves having been left aside for the time being. It was only in the early 1930s that scientists' attentions shifted from the structure of atoms to that of the atomic nuclei themselves.

Having been exposed to numerous drawings of what has become a standard 'sketch' of atoms, we have cultivated a mental picture of an atom that consists of a dot at the center — the nucleus — surrounded by several concentric circles, the orbits of electrons. The relative sizes of an atom and its nucleus are, however, vastly different, and it will take some numerical comparisons to drive this point home. On the average, atomic nuclei are some 50,000 to 100,000 times smaller that atoms. Let's take the carbon atom as an example. A carbon atom is about 3 angstroms across, that is, about three-tenths of a nanometer across while a carbon nucleus checks in at about a five-millionth of a nanometer across, that is, some 60,000 times smaller than the atom. There is indeed a vast emptiness between the nucleus and the electrons whirling around far away from it. We can put this 60,000-to-1 ratio in at least three different perspectives, in ascending order of magnitudes.

First, let us blow up the scale of a carbon nucleus to the size of a ball with a 1-foot diameter. A carbon atom, in this scale, would then be a sphere that is about 11 miles (or 60,000 feet) across. Imagine a giant modern jetport that is 11 miles across. Right smack in the center of the airport, in the middle of a runway, sits a basketball. A basketball and an airport ... now that is a lot of open space! Let's try another analogy, this time a little bigger in magnitude. The radius of the Earth is about 4,000 miles. Divide that by 60,000 and we get about 110 yards, goalpost to goalpost of a football field. Let us imagine that, on one cataclysmic Monday morning, the Earth suddenly collapses into itself! The whole Earth collapses into a ball — very dense and very hot — about two football fields in diameter, from the North to the South Pole! Let us imagine further that while the Earth collapses, the canopy of the atmosphere stays intact right where it is. In other words, the atmosphere stays put while the bottom drops out of the ground, the whole Earth condensing into a ball at its center. The nucleus — the collapsed Earth — is but a tiny little sphere occupying the dead center of the atmosphere.

We can cook up another analogy, this time going all the way to the dimensions of the solar system. The radius of the Sun is about 700,000 kilometers, while Pluto hangs around about six billion kilometers away from it. The ratio of the distances is 'only' about 10,000-to-1. The electrons of a carbon atom are swirling around the carbon nucleus at a distance that is six times further out than Pluto is from the Sun. A 'tenth' planet, six times further out than Pluto, would be to the Sun what an electron is to a carbon nucleus.

Protons and Neutrons: From A to Z

A proton is another name for the nucleus of a hydrogen atom. As an archetypal example of the planetary model of an atom, the proton accounts for slightly more than 99.95% of the mass of a hydrogen atom, weighing some 1,836 times more and being at least 1,000 times larger than an electron. Compared to an electron, a proton is a sheer brute. With humbling disparity, however, the electric charge of a proton only just balances out the negative charge of an electron, exactly to its last decimal point. When it comes to electric charges, the proton is an equal match to the electron, and no more. So much heavier and larger, yet a mere equal in charge. The number of protons — that is, the number of positive units of charge — is defined as the atomic number, denoted by its standard notation Z. The atomic number signifies hence either the number of protons or, just as equivalently, the number of electrons in an atom in its natural state of electrical neutrality. Z for the hydrogen atom is one, two for the helium atom, six for the carbon atom, and so on.

Insofar as the discovery of the atomic nucleus in 1911 was a generic one — that is, that it applies to all atoms — it certainly includes the nucleus of the hydrogen atom as well. The specific name, proton — designating the nucleus of the hydrogen atom — was not christened until 1914 by Rutherford himself. In that sense, we usually peg the year of discovery of the proton at 1914. Soon thereafter, another fundamental property of the proton was established, making it the literal equal of the electron. Determination of the nuclear spin showed that a proton had exactly the same amount of spin as an electron — that is, $1/2$ in our now-familiar encrypted notation — and as such, other than its massiveness, the proton turned out to be an equal match to the electron in both its charge and spin.

Now, if it had turned out that an atomic nucleus is an aggregate of only one kind of constituent, the protons, the systematics of all elements in the Universe could not have been simpler — 'solar' systems of uncompromising simplicity. One proton and one electron constitute a hydrogen atom, two protons and two electrons a helium atom, three apiece a lithium atom, and so on. Ignoring the mass of the orbiting electrons — in comparison to the 'ultraheavy' protons — the masses of elements would be simpler than a nursery rhyme. One, two, three ... for the hydrogen, helium, lithium ... atoms, in units of the mass of a hydrogen atom. A carbon atom would have 6 protons and 6 electrons, and it would weigh exactly 6 times as much as a hydrogen atom.

Well, things did not turn out to be that simple. A helium atom weighs some 4 times as much as a hydrogen atom, a lithium atom 7 times, and a carbon atom 12 times. Clearly there had to be something other than protons inside nuclei that is responsible for the 'extra' mass, but which is not carrying any electric charge; the matching numbers of protons and electrons neatly canceled out the net electric charges for atoms, and there was no way that any new constitutents could have been carrying net charges. What fitted this bill was the 'neutral proton,' something that had the same mass as a proton but which was not electrically charged at all. The existence of such a particle had been suspected, in fact, ever since the atomic nucleus was discovered; Rutherford is said to have conceived of such a 'neutral proton' as a hydrogen atom in which the electron had fallen into the proton.

It wasn't until 1932, however, that the evidence for such a particle was finally uncovered by James Chadwick (England, 1891–1974), who promptly named it the neutron. Its mass was virtually identical to that of a proton, 939 vs. 938 in terms of the now-familiar MeV over c^2, about one-tenth of 1% heavier. Later on, the spin of the neutron was also confirmed to be identical to that of the proton; all three — electron, proton and neutron — have an identical amount of spin. The total number of protons and neutrons that make up a nucleus is defined as its mass number, denoted by the letter A; while the total number of protons has its own designation, the atomic number Z. The total number of neutrons has no special designation of its own, other than the obvious difference given by A − Z. A list of properties for protons and neutrons looks like this:

Name	Proton	Neutron
Symbol	p	n
Charge	+1	0
Mass	938.3 MeV/c^2	939.6 MeV/c^2
Discovered	1914	1932
Spin	1/2	1/2

The fact that protons and neutrons are so much alike — the same mass, size and spin, differing from each other only in their charges — prompted a common generic name, the nucleon, designated by a new symbol N. It is often more convenient to simply lump protons and neutrons together and refer to them as nucleons, the proton being the positively-charged nucleon, N^+, and the neutron the neutral one, N^0. A new name, nucleonics, was coined later for technologies based on nuclear science, in a similar vein to electronics or photonics. The mass number A — the total number of protons and neutrons in a given nucleus — is then simply the total number of its nucleonic constituents.

Tight Quarters for Nucleons

The fact that the constituents of a nucleus — the charged and uncharged nucleons — are of the same mass immediately points to one of the many stark differences between the structure of an atom and that of its nucleus. Unlike the case of an atom with its heavy center and the orbiting lightweight electrons, an atomic nucleus has no fixed center defined by anything heavier than the nucleons themselves; the 'atomic sun' has no 'sun' of its own. A nucleus is instead a quite 'democratic' aggregate of equal partners — no center and no orbits around a center — just a bunch of equal weights hanging around together. The simplest nucleus is the nucleus of a hydrogen atom, namely, a proton itself. A deuteron — the nucleus of a deuterium atom — consists of one proton and one neutron, each going around the other, something like a double-star system in which two suns push and pull each other.

As we go up the chart to heavier nuclei, containing tens and then hundreds of nucleons, another stark difference in the structures between atoms and atomic nuclei comes into view. In sharp contrast to the case of atoms — which are full of the 'void' between the nuclei and the outlying electrons — the interior of a nucleus

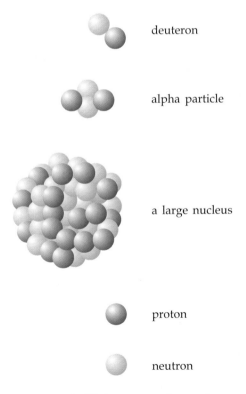

Figure 2 Tight quarters for nucleons.

is generally very crowded, with not much 'emptiness' to spare. Take the well-known example of the uranium nucleus; the most abundant type of naturally-occurring uranium is uranium-238, containing 92 protons and 146 neutrons. Despite its high mass number — highest among the naturally-occurring elements at $A = 238$ — the uranium nucleus is only about 14 protons across, from pole to pole. What we are looking at is a basketball stuffed with 238 ping-pong balls. It is fairly crowded place with not much empty space. If an atom is a 'planetary' system, a nucleus is a tightly bunched stem of grapes in two tones — green grapes for neutrons and purple grapes for protons!

Nuclei are often also referred to as isotopes — a table of isotopes, radioactive isotopes, and so on. Since the chemical properties of an atom are completely characterized by the number of electrons it contains, different nuclei that have the same number of protons — while having different masses — exhibit identical chemical properties, and for this reason the elements that differ from each other only by the number of neutrons in their nuclei are lumped together as isotopes. Hydrogen has two isotopes; deuterium with its nucleus deuteron — one proton and one neutron — and tritium with its nucleus triton — one proton and two neutrons. Helium, whose nucleus has two protons and two neutrons, has one well-known isotope — helium-3, whose nucleus has one neutron less than the standard helium-4. The mass number for both tritium and helium-3 is 3, but the atomic number is 1 for the former (one proton and two neutrons) and 2 for the latter (two protons and one neutron).

A Brave New World of Tempest Within

So, what at first appeared to be a simple dot at the center of an atom turned out to be something quite extraordinary, an exceedingly small area — some 100,000 times smaller than an atom — stuffed with a bunch of nucleons that were virtually identical to each other, except some were positively charged (protons) and others not (neutrons).

The most immediate logical question that raises itself is, what holds this bunch of nucleons tightly glued together? What is the nature of the force that operates among the nucleons making up the atomic nuclei? Two things about this force jump out at you immediately: for one thing, it is clearly not electric force — the force that forms atoms, holding electrons firmly attached to the atomic nucleus. It is too obvious. Neutrons carry no electric charge and whatever force is responsible for holding neutrons in place clearly does not depend on electric charges. The second obvious thing about the force shows itself with respect to the binding of protons: not only is the new force independent of charges, but it is clearly strong enough to overcome them. Since protons are all positively charged, the electric forces among the protons are all repulsive — like charges push off against each other — so whatever force is holding protons together (they don't fly apart) has to be an attractive force that is stronger than the electric repulsion between the protons. Stronger than electric force, attractive, and independent of charges. Clearly, this had to be an entirely new kind of force, the likes of which had not been encountered before.

What is even more strange about this force is that it completely disappears, without a trace, outside the walls of nuclei. The atomic electrons orbiting around a nucleus are totally unaware and unaffected by it; the new force operates and exists only within the extremely reduced dimensions of atomic nuclei. It is a force that is severely limited in its effective range of operation. It was first named the nuclear force, for an obvious reason; but in the course of several decades following the early 1930s the name for this new force went through some changes. Honoring its strength, the word 'strong' was added, making it the strong nuclear force, and later still the word 'nuclear' was dropped altogether, making it just the strong force. In some quarters it is also called the strong interaction. One reason for dropping the designation 'nuclear' from the strong force was the realization that there was, hidden within the walls of atomic nuclei, yet another new type of 'nuclear' force entirely distinct from the first, and it soon became somewhat redundant to refer to both of these new forces as being 'nuclear.'

Within the walls of nuclei, there operated another force — extremely feeble but distinctly different — that affected the nucleons, a second kind of the 'nuclear force.' Every once in a while — not often but often enough to notice — the protons and neutrons would change into each other! Yes, a neutron would turn into a proton and vice versa. When this happens, the mass number A — the total number of the nucleons — would not change, but the number of protons — the atomic number Z — and accordingly the number of neutrons, would each change by one. The same stem of two-tone grapes, but every once in a while one green grape would turn purple or one purple green!

Whatever force was responsible for this little trick, it wasn't the same force holding nucleons together; it was something else altogether. It would be a while, until the 1950s, that it was gradually realized that we were dealing with another kind of 'nuclear force' from the first. One characteristic of this 'second nuclear force' was its extreme feebleness; in comparison with the strong force, and even in comparison with the traditional electric force, this 'second force' was unspeakably weak, a mere whisper in the midst of a tornado. This feeble new force was first named the weak nuclear force, but was soon thereafter renamed the weak force, or the weak interaction.

The science of atomic nuclei is thus a study of not one but two entirely new forces of nature; indeed a brave new world. For reasons scientific as well as historic, the notions of atoms and atomic nuclei are usually lumped together in the minds of the general public. After all, we are talking about atoms and their structures, only in varying degrees of detail. From the scientific point of view, however, this is an inaccurate picture. It is important to realize that the two sciences — that of atoms on the one hand and that of atomic nuclei on the other — are fundamentally different branches of physics. In the case of atomic physics we are dealing with the consequences of the electric force that binds electrons to their nucleus, and in the case of nuclear physics we are studying the nuts and bolts of two entirely new forces of nature — the strong and weak forces. It is a whole new ball game — new playing field, new players, new rules and new everything.

7

The Strong Force I: Nucleons

The physical differences between the 'old' forces — the electric force and gravity — and the 'new' — the two new types of forces inside atomic nuclei — are quite drastic. None of the concepts that we are familiar with (such as mass, weight, electric charge, electric current, or the well-known inverse-square law) have any place in the scheme of the new forces. The most immediate and drastic difference lies in the fact that the 'new' forces exist only within the confines of atomic nuclei; once outside they disappear without trace! Bearing in mind that the average nucleus is some 60,000 to 100,000 times smaller than an atom, the distance over which the new forces exist is, by our standards, virtually equal to zero. In most cases, the reach of the new forces is even shorter than the diameter of a nucleus; a nucleon at the 'south pole' of a nucleus does not feel the new forces from one located at its 'north pole.' Compared with the 'old forces' of gravity and electricity, which extend their influence from one end of the Universe to the other, the new forces are virtually 'point-contact' forces — their range of influence extends no further than a point! It is thus clearly impossible for us to devise any tabletop experiment in which we can directly observe the effects of these new forces, and that makes it a bit difficult for us to develop an intuitive feel about them. We will talk about the weak force in the next chapter; here we will focus on the first of the two 'nuclear' forces, the strong force.

The Fury of a Twister

The strong force that keeps nucleons — protons and neutrons — held together tightly to form atomic nuclei lives up to its name; it compensates for a short range of influence through its Herculean might. Pound for pound, the strong force is determined to be some 140 times more powerful than the electric force. The technique for measuring the strength of the strong force is not as simple as those for measuring the gale force of a storm or the stiffness of a heavy coil spring, but the idea is the same. Suppose we are given a heavy coil spring that requires an engine of 140 horsepower to compress an inch. The electric repulsion between a pair of protons is like a smaller engine of only 1 horsepower. The spring simply would not budge. The relative strengths of the strong and the electric forces are like that, 140-to-1. For a pair of neutrons or a pair consisting of a proton and a neutron, no electric force operates between them and a relative strength of 140 of the strong force is all that exists between them. The best place to measure and 'feel' the brute strength of the strong force is in a deuteron — the nucleus of the deuterium (that is, the heavy hydrogen). A deuteron is a light nucleus made up of one proton and one neutron. The amount of energy, or power, needed to break a deuteron apart into a free and unbound proton and neutron gives us a good measure for the strength of the strong force. Compared to the 'meager' electric force, the strong force is a 'heavy coil spring'; it is like comparing a suspension spring for a military tank to the small spring inside a ballpoint pen!

The strong force, with its strength and limited range, has no ready equal in our everyday world. One analogy that comes close to bringing out both the strength and the range is the fury of a twister; the ferocious destructive power of a tornado. No need to elaborate upon the strength and power of a twister, which destroys and flattens everything in its path. But — and here is the analogy — the range of a twister, the lateral width of its path, is often no more than a few city blocks across. A block away from the cylindrical path of a twister, a building may suffer only a few broken windows. Powerful, yes, but very narrow in its breadth. Some 140 times more powerful than the electric force, but working only within the confines of an atomic nucleus, the strong force is a strange new breed of force.

It is the sheer power of this force hidden inside nuclei that when unleashed represents the awesome power of the nuclear age — nuclear reactors, nuclear power, nuclear fission, nuclear fusion, and all that. From the moment atomic nuclei were first formed in the evolution of the Universe, this brute force was carefully packaged and hidden inside them, and only in the 20th century have we come to know this age-old secret.

The Nucleonic Charge

By the nature of its confined range — being effective only within the dimensions of atomic nuclei — there is no way that the strong force can affect the electrons that orbit at the 'outermost' reaches of an atom. It turns out that the strong force does not at all affect electrons, positrons or photons under any circumstances. Since the early 1970s, scientists have been able to shoot electrons and positrons deep inside atomic nuclei, hurling them toward protons and neutrons in a close 'fly-by,' coming well within the range of the strong force, but the results show no response at all; as far as electrons, positrons and photons are concerned, the strong force might as well not exist!

This aspect of the strong force is, however, nothing strange; different forces have different 'clients,' so to speak. The good old electric force affects only those that carry electric charge, either positive or negative; it has no hold on any electrically neutral objects. A photon, or, for that matter, a piece of paper does not respond at all to the push and pull of electric force; they are electrically neutral. This simple parallel was also observed in many subsequent new types of charges.

If electrons, positrons and photons are totally impervious to the pull of the strong force, what could be simpler than invoking a concept of a 'nucleonic charge,' and saying that these 'lightweight' particles are 'nucleonically neutral.' The protons and neutrons — the nucleons — would then carry, by our own assignment, a positive one unit of this new nucleon charge; a proton has the electric charge +1 and the nucleonic charge +1, while a neutron has the electric charge 0 and the nucleonic charge +1. The electrons, positrons and photons? In addition to their usual electric charges, −1, +1, and 0, respectively, they all carry zero nucleonic charge. The cryptic list of properties of the nucleons thus picks up another entry, the nucleonic charge:

Name	Proton	Neutron
Symbol	p	n
Mass	938.3 MeV/c^2	939.6 MeV/c^2
Discovered	1914	1932
Spin	1/2	1/2
Electric Charge	+1	0
Nucleonic Charge	+1	+1

Expressed in numerical terms, as in +1 or 0, the electric and nucleonic charges look deceptively similar, but there is one qualitative difference that may be worth noting here; in a nutshell, while as far as the electric charges are concerned, the numerical designation is a matter of convenience, in the case of the nucleonic charge it is more than this. In the case of the electric charges, the scale for the charges are established by the inverse-square law; the charges can be and are in fact expressed in terms of well-known, readily measurable quantities such as force, distance, amperes and so on. The numerical designation +1, for the electric charge, is a shorthand for $+1.6 \times 10^{-19}$ coulombs, the flow of one coulomb of charge in one second being equal to one ampere. That amount, that is, 1.6×10^{-19} coulombs, was the 'quantum' for all charge transfers, and designating it as one was a matter of convenience, the charge of an electron being denoted as −1.

In the case of the nucleonic charges, the numerical designation +1 is not a shorthand for any other quantity, for one very simple and frustrating reason; after some seven decades of studying it since the early 1930s, despite a mountain of knowledge we have gained of it, to this date we are unable to write down the force law for the strong force. The strong force, with all its peculiarities — the extremely short range of it being just one example — has to this date defied all attempts to formulate a definitive expression for its law. We have the inverse-square law for the force of gravity, and similarly we have the inverse-square law for the electric force. We do not — repeat, NOT — yet have an expression for the law of the strong force. Not that we are totally in the dark; over the decades we have accumulated a huge amount of information about the strong force and been able to come up with some reasonably approximate expressions, but only with limited applicability. We have yet to be able to write down a definitive law for it. And in the absence of a rigorous rule for the force, the amount of the nucleonic charge that we have somewhat arbitrarily assigned cannot be expressed in terms of the familiar units we use for forces, distances, or any other such quantities. For this reason, the designation +1 for the nucleonic charge remains just that, +1; it is not a shorthand for any other numerical quantity.

This practice of assigning numerical 'charges,' in complete emulation of the electric charges, has in fact come to dominate the world of nuclear and particle physics. Assignments are carefully made in such a way that for each newly-invoked 'charge' there corresponds its own 'zero-sum rule,' its own law of conservation. The 'ritual' of invoking new charges along with their associated zero-sum rules constitutes in fact the very foundation upon which is built our present-day knowledge of nuclear and particle physics.

Antinucleons

Since the inaugural discovery of antimatter — the discovery of the positron in 1932 — it was more a question of when than if antimatter counterparts of other particles such as the proton and the neutron would be discovered. Even so, the actual confirmation of this went a long way toward firmly establishing the matter-antimatter symmetry as one of the new truths of 20th century physics. A proton should have its mirror image, the antiproton: a particle that would have the same mass and spin as a proton but with the opposite (that is, negative) electric charge. Just such a particle was indeed observed in 1955 at a particle accelerator located in Berkeley, California. Bearing in mind that a proton weighs some 2,000 times as much as an electron, a very highly energetic photon — having energy some 4,000 times greater than the ones that can create an electron-positron pair — had to become available inside a man-made particle accelerator. These photons then went through the same dance of matter-antimatter creation, and instead of producing a pair consisting of an electron and a positron, they created a pair consisting of a proton and an antiproton. This is how in 1995, the unmistakable tracks of antiprotons were identified by a team of physicists working at a facility in Berkeley. The standard notation for the antiproton is $\bar{p}$, that is, p with a bar on its top, pronounced p-bar.

It is when we try to make a correct assignment of the nucleonic charge to an antiproton that we come across the first emulative

The Strong Force I: Nucleons

extension of the concept of the electric charge: just as the electric charge of an antiparticle is opposite from that of its corresponding particle, we assign the value −1 for the nucleonic charge of an antiproton, that is, all 'charges' of an antiproton are opposite those of a proton. The matter-antimatter symmetry now has two charges — electric as well as nucleonic — that define all mirror images. The property list of an antiproton is therefore:

Name	Antiproton
Symbol	$\bar{p}$
Mass	938.3 MeV/c²
Spin	1/2
Electric Charge	−1
Nucleonic Charge	−1

Now, insofar as both protons and neutrons are identical in the eyes of the strong force, if a proton has an antiproton, then a neutron should have its own anti, namely, the antineutron: a particle that would have the same mass and spin. A neutron being electrically neutral, however, the antineutron would also have to be electrically neutral. A natural question here is, in what sense is an antineutron the 'anti' of a neutron?

Having invoked the new nucleonic charge and assigned the value +1 to both protons and neutrons ('identical' in the eyes of the strong force), this leads naturally to the assignment of −1 to the antinucleons. An antiproton has not only its electric charge but its nucleonic charge as well, opposite to that of a proton, so an antineutron is the anti of a neutron in the sense of its having an opposite nucleonic charge from that of a neutron. The implication of the invocation of the nucleonic charge, and of assigning opposing values of it to nucleons and antinucleons, cuts deeper than meets the eye; it has extended the definition of antimatter from just having opposing electric charges to all 'charges'— electric as well as nucleonic. This extension recurs several times in the course of the development of particle physics; each time a new 'charge' is invoked, it adds one more 'charge' by which an antiparticle is opposite from its corresponding particle. The list for an antineutron (n-bar) is thus:

Name	Antineutron
Symbol	$\bar{n}$
Mass	939.6 MeV/c²
Spin	1/2
Electric Charge	0
Nucleonic Charge	−1

An antineutron is the anti of a neutron by virtue of the opposite nucleonic charges, but both are electrically neutral. An aggregate of antiprotons and antineutrons would constitute an antinucleus. A system of antielectrons — positrons — orbiting around an antinucleus would be an antiatom. An antihydrogen atom consisting of an antiproton at its center and a lone positron going around it would be just as electrically neutral as the hydrogen atom, but they would clearly be the anti of each other. With antiatoms, we could then form antimolecules, antiwater, antibeer, anticheese, and so on up to entire antiworlds! There are no intrinsic differences between a world and an antiworld, except for one thing: if they met up with each other, there would be a cataclysmal annihilation in one horrendous 'bang!'

The Nucleonic Zero-Sum Rule

The introduction of the concept of the nucleonic charge, with its equal and opposite assignments to nucleons and antinucleons, gains its meaning and status only when it is accompanied by a zero-sum rule all of its own. It is then that the idea becomes as fundamental as the electric charge and its zero-sum rule, and the validity of the conservation of the nucleonic charge has since been established beyond any question. Of all reactions involving nucleons by themselves, the zero-sum rule of the nucleonic charge has not to this date seen a single deviation.

Suppose we have a reaction in which a highly energetic proton — traveling at almost the speed of light — smashes into a neutron. Usually this is done by accelerating protons inside a high-energy particle accelerator — which pushes protons up to speeds close to the ultimate speed of nature (the speed of light) — and then steering them smack into a sitting deuterium nucleus, consisting of one proton and one neutron. Half of the reactions will be between protons and protons and the other half between protons and neutrons. In a proton-neutron smash-up, the electric charge of the initial system is +1 (+1 from the proton and 0 from the neutron) and the nucleonic charge is +2 (+1 each from proton and neutron).

After the violent smash-up the available total energy of the system converts itself ($E = mc^2$ rides again) in different channels of several new nucleons; a whole new bunch of protons, neutrons, antiprotons and antineutrons. Of all manner of mathematical possibilities, only those whose total charges match the initial condition — the electric and nucleonic charges being +1 and +2 respectively — can be realized, all others are forbidden and do not occur. Those that violate the zero-sum rules will not occur, and include such final configurations as two protons and a neutron (+2 and +3, respectively), two protons and an antineutron (+2, but +1), or a neutron, an antineutron and an antiproton (−1 and −1). A final product such as two protons, one neutron and one antiproton (+1 and +2) is certainly allowed.

The nucleonic charge and its conservation law represent the first successful extension of the concept of the electric charge and its conservation law into the domain of the subnuclear world of elementary particles, where more often than not we are much less sure of things than with the macroworld of electricity and gravity. The invocation of the nucleonic charge was in fact only the first of a series of such extensions, as we shall soon see. Throughout the 1950s and 1960s, we came to discover hordes of new particles, associated in one way or another with nucleons. Some of these newly-discovered particles were so much like nucleons, they necessitated extending the idea of nucleons to a much larger group of particles, called baryons. We will discuss them in a later chapter, but for now we turn our attention to the weak force, the other 'nuclear' force that operates exclusively within the confines of the dimensions of atomic nuclei.

8

The Weak Force: A Whisper in the Night

The discovery of the strong force — a new force of nature concealed deep inside atomic nuclei — is definitely one of the most significant milestones of 20th century science. It ushered in the nuclear age, the age of nuclear energy and power — nuclear reactors, nuclear power plants, nuclear propulsion systems, and nuclear medicine, not to mention nuclear weaponry. Yet, to this day, after some six decades after its initial discovery, we are not as familiar with the strong force as we are with the other two forces of nature — the gravitational and the electric forces. The very fact that the strong force stays packed within the confines of atomic nuclei certainly does not make it easy for us to study it directly, and this has resulted in a somewhat embarrassing state-of-the-affair, in which we have not yet been able to clinch a concrete expression for its force law. Nothing as simple and elegant as the inverse-square law exists for the strong force.

As if the riddle of the strong force isn't baffling enough, the interior of the atomic nucleus harbors, it turns out, yet another mystery, much deeper and even more baffling; there is yet another entirely new type of force that lurks within, playing tricks on neutrons and protons. Every once in a while, neutrons and protons inside atomic nuclei spontaneously change suits — a neutron becomes a proton and, a little less often, a proton turns into a neutron — a dance of changing suits that nevertheless keeps the total number of nucleons (hence the mass number A) of a nucleus unchanged.

When this changing of suits take place, it happens in strict adherence to the zero-sum rules of both the electric and the nucleonic charges. The conservation of the nucleonic charge is obvious enough: as a neutron changes into a proton, or vice versa, there is no net change in the nucleonic charge — it remains the same, +1 before and +1 after. The adherence to the strict zero-sum rule of the electric charges is accomplished by creating, at the moment of the transmutation, either an electron or a positron. As a neutron, with zero electric charge, turns into a proton, an electron is created and emitted on the spot, neatly canceling out the positive charge of the proton. The converse process — a proton turning into a neutron — is accompanied by the creation and emission of a positron, the positive charge of the antielectron matching the positive charge of the proton. The two zero-sum rules are as exacting as they are inflexible.

A Whisper in the Night

The way we observe this flip-flop between protons and neutrons is by the spontaneous transmutations among related atomic nuclei. In a typical process, a radioactive carbon isotope called carbon-14 turns itself into a nitrogen-14 nucleus with the accompanying emission of an electron. The former has 6 protons and 8 neutrons, the latter 7 protons and 7 neutrons. One of the eight neutrons of the carbon-14 nucleus apparently changes into a proton and the newly-created electron is ejected out of the nucleus. There is a converse process also; a nitrogen-12 nucleus with 7 protons and 5 neutrons switches into a carbon-12 nucleus, with 6 protons and 6 neutrons. This time it is not an electron but rather its antiparticle

— the positron — that is ejected from the nitrogen-12 nucleus. The latter process, and those similar to it, is in fact the main source for collecting a bunch of positrons for the purpose of experiments on antimatter, as well as medical imaging applications such as the imaging technology called PET, positron emission tomography.

In the scale of nuclear reactions, this new class of nuclear transmutations occur very infrequently and with very small amounts of energy involved. In a world dominated by the Herculean strength of the strong force — where things change in a matter of a billionth of a billionth of a second — these effects take place in a matter of 'only' a billionth of a second; relatively speaking, once in a blue moon! Whatever mechanism is responsible for this changing of suits between protons and neutrons, it is a very feeble one; a barely audible whisper in the midst of a hurricane. However weak and feeble, however, there is no mistaking its signature; a carbon-14 turns into a nitrogen-14 and a nitrogen-12 changes into a carbon-12 in every so many billionths of a second, every single day.

It took much guesswork and keen insight before Enrico Fermi (Italy, USA, 1901–1954) first put forward an idea that this may be the work of an entirely new kind of force operating deep inside atomic nuclei. If it was indeed due to a new force, the force was a very weak one. From the relative 'slowness' of its effect, one could surmise that the strength of this new force would have to be at least a million times weaker than the strong force. It was a pretty weak force, all right. The proposed new force was first dubbed 'Fermi interaction,' something of a temporary designation. In time, the name would go through its own stages of evolution — the weak interaction, the weak nuclear force, and to its current name, the weak force. The weak force is thus the fourth force of nature, after gravity, electricity, and the strong force. In one century, therefore, we had the discovery of not one but two entirely new types of force.

The weak force, while sharing some common characteristics with the strong force, displays some entirely unique traits of its own. Both the strong and the weak forces are restricted completely to the confines of atomic nuclei; they simply do not exist beyond the nuclear dimension. The atomic electrons orbiting around their nucleus couldn't care less about the two new forces; they do not feel their respective tugs at all. The strength of the weak force is so feeble, in fact, that unlike the case of other forces, we have not been able to find any system that is held together, for any respectable duration, by the weak force alone. The atoms are held together by the electric force between the orbiting electrons and the positive electric charges of the protons inside the nucleus, and the protons and neutrons are glued together by the strong force, making up atomic nuclei. But no system has been observed to date that is held together by the effects of the weak force alone; it is just too feeble.

There is one more remarkable aspect that differentiates the weak force from the strong forces. As we have stated, when a neutron turns itself into a proton, the process also creates and emits an electron, and conversely when a proton inside a nucleus changes into a neutron, a positron is created and emitted. Apparently the weak force involves electrons and positrons as well as nucleons; the strong force that works on the nucleons is totally oblivious to electrons and positrons. The weak force affects both nucleons and electrons, as well as their antimatter counterparts, positrons and antinucleons. A whisper in the night in the midst of a hurricane, yes, but it is nevertheless heard by all. So feeble, so well hidden inside atomic nuclei, and piggybacking on top of a force that is at least a million times stronger. The story of the weak force is, by any measure, a strange one, and it becomes even stranger.

The Neutrino: The Electron's 'Weak Shadow'

As we have noted, the process of transmutation of a neutron into a proton, and likewise a proton into a neutron, proceeds in strict adherence to the zero-sum rules of the electric and nucleonic charges, respectively. The nucleonic charge of +1 remains unchanged before (a neutron) and after (a proton and an electron), and, similarly, the electric charge of 0 remains unchanged before (a neutron) and after (a proton and an electron). Things seemed nice and simple, at this stage. However, there remained one persistent and nagging imbalance — no matter how carefully one measured the process, the energies of the resultant proton and electron did not fully add up to the initial energy of the parent neutron. The energy-mass equivalence relationship of $E = mc^2$ was fully taken into account, but each and every time the total energy of the end product — a proton and an electron — always turned out to be just a tad less than the energy put into the system, that of the parent neutron. Not much, but a tiny bit of energy was missing and unaccounted for. You started out with one dollar and ended up with only 99 cents; something was carrying off a penny.

Since the zero-sum rules for the nucleonic and electric charges were all precisely checked out, whatever was draining a tiny bit of energy in the course of the transmutation could not have been anything that carried either a nucleonic or an electric charge. The mystery of the missing energy deepened when it was determined that no other trace of anything with mass was involved. In other words, no other particles in the conventional sense — having some mass and carrying some electric charge — were involved. Whatever it was, if it was anything, it had neither mass nor charge. If it was to be a particle of some kind, it had to be a massless and chargeless one.

Now, to be sure, an idea of a massless and chargeless particle was not something new: the concept of the photon — the quantum of electromagnetic radiation — accommodated such properties. But by the time the case of missing energy came to be noticed, in the late 1920s, the concept of the photon had been well established for a little over two decades. Besides, the technique for detecting the presence of photons had been used to decide whether the missing energy was a photon or not. It wasn't. Whatever was responsible for this imbalance, no more than a few percent off, was not at all related to electromagnetic radiation, although the two hallmark characteristics of the photon were definitely shared.

Viewed as a particle— carrying off a little bit of energy but with neither mass nor electric charge — that is created and emitted in unison with the product electron, it came under the scrutiny of yet another zero-sum rule, that of the spin. Remember that all the particles we have met so far — protons, neutrons, antiprotons, antineutrons, electrons and positrons — all have the same and identical amount of spin as an electron; in the parlance of quantum physics, they all have spin one-half. The sole exception to this is the photon. A photon is assigned a spin that is twice that of an electron; a spin of one. When the zero-sum rule for the spin of the involved particles was invoked — the spin of the neutron must equal the sum of the spins of the proton, electron and the 'missing' particle; it became clear that the 'missing' particle must carry a spin that is also identical to that of an electron. (Spins do not add like numbers. Just as a clockwise rotation would be canceled out by a counterclockwise rotation, the spins bring into their sum factors related to their direction.) Seen in this light, it was definitely not a photon; its spin was off by a factor of two. It had to be a new particle that had no mass, no electric charge, but managed to carry some energy and spin in the amount identical to that of an electron.

Not surprisingly, when the idea for such a new phantom particle was first proposed in 1931 by Wolfgang Pauli, skeptics abounded, with chuckles here and sneers there. Enrico Fermi was the first to realize the value of Pauli's idea. The idea of just such a particle was essential to Fermi, who was developing, for the first time, some theoretical framework to explain the weak force of nuclear transmutation. It fit his picture, and he gave it its name, the neutrino. A neutrino is thus a neutral partner of an electron, in a manner analogous to the neutron's relationship to a proton; it accompanies the electron in each and every process initiated by the weak force. As far as the work of the weak force is concerned, the neutrino shadows the electron at every turn; it is always there. As a neutron changes, within an atomic nucleus, into a proton, the process creates and emits not one (an electron) but in fact a pair of featherweights — an electron and its neutral 'shadow,' the massless, chargeless, and hence, formless phantom dubbed the neutrino. This association — the pair of nucleons, the proton and the neutron, on the one hand, and the pair of featherweights, the electron and the neutrino, on the other hand — is a remarkable parallel that persists to this day and in fact, coming into the 1970s, formed the founding pattern for our current understanding of matter.

The history of the neutrino is a study in sparsity. As we said earlier, when Wolfgang Pauli audaciously proposed the idea in 1931 it was born out of a desperate need to balance out the energy. When Enrico Fermi submitted his paper in which he outlined a theory for the weak force, fully incorporating the particle Pauli had proposed — which Fermi christened the neutrino — the paper was summarily rejected by the editor of the journal *Nature*! The idea remained just that, a theoretical construct, for almost a quarter of a century, until 1955. That year, for the first time, telltale signs of the feeble phantom of a particle were detected among the products of a nuclear reactor facility in Savannah, Georgia by a team led by Frederic Reines (USA, 1918-1998) and Clyde Cowan (USA, 1919-1974). It would be another 40 years until 1995, when Reines was recognized for his work with a share of the 1995 Nobel Prize in Physics. By then he was in a poor health and Cowan had long before passed away, in 1974. All told, it took some 64 years, from 1931 to 1995, for the neutrino to claim its rightful place in Nobeldom.

Leptonic Charges for the Leptons

The symbolic parallelism between the proton and the neutron on the one hand, and the electron and the neutrino on the other, is inescapable. The proton and neutron form a pair — we refer to them indistinguishably as nucleons — that behave in an identical manner not only with respect to the strong force, but in fact also with respect to the weak force. By virtue of the latter, they routinely change suits with each other. Other than a tiny bit of mass difference of about 1 MeV/c^2–939 MeV/c^2 for a neutron vs. 938 MeV/c^2 for a proton — and the obvious difference in their electric charges — they are for all intents and purposes identical particles in the eyes of not only the strong force but also the weak force.

The mass difference between the lighter pair, the electron and the neutrino, is also of the same order; since a neutrino is massless the difference is all of the electron's mass — about 0.5 MeV/c^2. Neither the electron nor the neutrino has anything to do with the strong force, but with respect to the weak force, both are — for all intents and purposes — identical; they participate in the actions of the weak force as equal partners. It is only too natural then to do with the electron and neutrino what we did with the proton and neutron; if the latter can be viewed as two different manifestations (by the electric charge) of one and the same particle (the nucleon) then we can likewise consider the electron and the neutrino to be two different faces (again by the electric charge) of one and the same particle. The name chosen for this single entity, the lepton, was a logical choice, from the Greek word 'lepto' meaning 'slight' or 'light.' The electron is the (negatively) charged lepton and the neutrino the neutral lepton, in an exact parallel to the pair of 'heavies' — the proton is the (positively) charged nucleon and the neutron the neutral nucleon. Thus, we speak of two pairs of particles, the heavy nucleons and the featherweight leptons. As far as the two new forces within atomic nuclei are concerned, the former enjoy the attention of both the strong and the weak, while the latter respond only to the feeble signal of the weak.

The nucleon-lepton parallelism can now be extended one step further. The idea of defining the nucleonic charge (+1 each for the proton and the neutron) for the nucleons can be extended to the leptons: the leptonic charge is defined in such a way that the leptons are assigned the value +1. The emulation and extension of the idea of the electric charge has thus gone one step further. Since the electron has its antiparticle, the positron, it stands to reason that the neutrino should also have its own antiparticle, the antineutrino. As a pair of antileptons, both the positron and antineutrino would be assigned −1 for their leptonic charge. The flip side of this is that the nucleons would have zero leptonic charge, and, similarly, the leptons would have zero nucleonic charge.

This extension — from nucleonic charge for the nucleons to leptonic charge for the leptons — may appear at first sight to be a straightforward parallel. In fact, it represents a bold conceptual leap. In the case of the electric and nucleonic charges, the invocation of the concept of charges rested on two fundamental *raisons d'être*: as the source of the respective force — the electric charge for the electric force and the nucleonic charge for the strong force — and at the same time as quantities that obeyed strictest zero-sum rules — the conservation of electric charges and of nucleonic charges for any reactions involving these elementary particles. The weak force, however, affects both nucleons and leptons. The assignment of the leptonic charge to leptons does not therefore represent any resemblance to the possible source of the weak force;

the act of invoking the leptonic charge rests entirely on how well such assignment is associated with yet another zero-sum rule, the conservation of the leptonic charges in elementary particle reactions.

With the neutrino balancing out the 'missing' energy, the transmutation of a neutron into a proton is accompanied by the creation and emission of the pair of leptons, the electron and the neutrino. The zero-sum rules for the electric and nucleonic charges are exactly satisfied, as we have already mentioned. How is the zero-sum rule of the leptonic charge applied here? Since the initial parent neutron has zero leptonic charge (it is not a lepton), the sum of the lepton charges for the final product — proton, electron and neutrino — must come out to be zero as well. In order for this to be satisfied, what must happen is that the 'neutrino' that accompanies the emitted electron has to be an antineutrino! It is by the rule of the leptonic charge zero-sum rule that the 'neutrino' produced in association with an electron has to be an antineutrino, and, conversely, the 'neutrino' that is produced in association with the positron — as a proton turns itself into a neutron, within a nucleus — is the 'regular' neutrino! The distinction

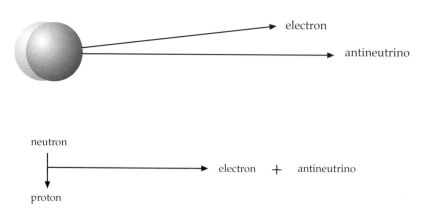

Figure 3 The weak force at work.

between a neutrino and an antineutrino is just this; one has the leptonic charge +1 and is paired with a positron, and the other has the leptonic charge −1 and is paired with an electron. The leptonic zero-sum rule is admittedly a bit artificial, compared to those for the previous two charges — electric and nucleonic — but so invoked, no violation of the rule has ever been observed to date.

The spreadsheet for the three types of charges now looks a little full:

	Electric Charge	Nucleonic Charge	Leptonic Charge
Nucleon			
Proton	+1	+1	0
Neutron	0	+1	0
Lepton			
Electron	−1	0	+1
Neutrino	0	0	+1

For the corresponding antiparticles, all charges reverse sign:

Antinucleon			
Antiproton	−1	−1	0
Antineutron	0	−1	0
Antilepton			
Positron	+1	0	−1
Antineutrino	0	0	−1

Two Pairs and Three Forces

The two pairs — a pair of 'heavyweight' nucleons and a pair of 'featherweight' leptons — represent a picture of the origin of matter that is as simple as it is powerful. Fully 99.99% of all known matter in the present-day Universe is made up from just the two pairs — nucleons and leptons — with the interlocking play among three forces of nature.

Under the influence of the mighty but short-ranged strong force, protons and neutrons stick together to form collective lumps, the atomic nuclei. Positively-charged nuclei become the centers around which move a set of orbiting negatively-charged electrons, forming atoms; atoms are held together by the attractive electric force between the electrons and nuclei. To make things a bit more interesting, there is this very feeble force hiding deep inside nuclei; this weak force performs the trick of turning protons and neutrons into each other, causing a nuclear transmutation, and creating in its wake a pair of leptons, either an electron and an antineutrino or a positron and a neutrino.

In the meantime, atoms attract other atoms by the electric force to form molecules, forming gases and liquids. Sometimes hundreds upon thousands of molecules coalesce into giant molecules. Some of these giant molecules take the shape of various solids while others become the molecules of life. As the chain becomes larger, we get flowers, trees, rocks, rivers, the planets, the stars, and ... the whole known Universe. And when the masses become significant enough, good old gravity comes into play. In the scale of the minuscule masses of the subatomic particles, the force of gravity can be completely left out of consideration.

For all practical intents and purposes, as far as the ordinary matter of the present-day Universe is concerned, the story of the origin of matter could end right here — two pairs and three forces. The physics of the elementary particle — and the contents of this book as well — could have stopped right here, and everyone could have lived happily thereafter. Alas, that is not how it was meant to be: there are yet more to come — more particles, more kinds of new 'charges,' and still more zero-sum rules, not to mention more chapters to follow in this book. It is perhaps appropriate, however, for us to take a brief pause here, with the pairs of nucleons and leptons as the 'end of the line' of the origin of matter, and wonder for a fleeting moment how simple it could all have been!

9

The Strong Force II: Hadrons

The first sign that things would not end with a simple picture of two pairs — one each of nucleons and leptons — came not from the leptonic but the nucleonic sector, the realm of the strong force. Almost as soon as the existence of the strong force was established, back in the early 1930s, a need arose for a handful of additional particles. At first, the name given to these new particles was the mesons, but, as with so many other names in elementary particle physics, it went through several stages of evolution, from the meson to the pi-meson, and then to its current title, simply, the pion (rhymes with high-on). The original name, meson, was retained to designate an extended family of pion-like particles, which were uncovered later.

Soon after the existence of pions was confirmed in the 1940s, a horde of new particles similar to nucleons and pions was uncovered over the next two decades. A new name was coined to denote these new particles: the strongly interacting particles, or SIPs, for they came about as a result of the strong force among nucleons and pions. Coming into the 1950s, with the advent of ever more powerful particle accelerators capable of generating ever greater amounts of energy, the number of new SIPs exploded exponentially, first by tens and eventually by hundreds. As the number of

SIPs increased, great effort was expended to sort out what at first seemed a bewildering proliferation of particles. Soon, some recurring patterns and systems of grouping emerged, and different groups of SIPs began to fit into definite slots, reflective of their relation-ships to one another. This, in turn, led to the suspicion that perhaps the SIPs — including protons and neutrons themselves — were not as elementary as had been thought, but rather composite structures made up from yet another layer of constituents of matter. The story of SIPs is first one of the great proliferation of particles, followed by the revelation of an underlying simplicity. First, we'll take a look at the pions.

Pions: The 'Strong Photons'

As soon as it was realized that the strong force is severely limited in its range of effectiveness, it became clear that the conventional definition of what a force is — one object influencing another over some extended distance — had to be drastically updated. A new vision of the microscopic mechanism by which one particle exerts force on the other was required. Fortunately, the guiding light was not far away. Remember how the electric charge and its associated zero-sum rule paved the way for the introduction of the nucleonic and leptonic charges, with their respective zero-sum rules? In a similar way, the key to understanding the mechanisms of the strong force, and eventually of the weak force as well, came from our knowledge of the electric force. That is, how electric charges interact with each other, or how an electron interacts with the quanta of electromagnetic radiation, the photons.

As we described in a previous chapter, the processes of absorption and emission of radiation by matter has its atomic explanation at the most fundamental level. Matter absorbs or emits radiation by means of individual atomic electrons either absorbing or emitting one photon at a time. This is how an atom interacts with light, by letting one of its electrons either gulp down or cough out a single photon at a time. The photons, as the particulate quanta of radiation, are the messengers that convey the electromagnetic field, and thus the electric force itself. Just as two basketball players race down the length of the court in a fast break, passing the ball back and forth, two electrically-charged particles — be they protons, electrons, positrons or antiprotons — play the subatomic 'give and go.' One particle emits a photon and the other catches it, back and forth in a continuous exchange of a stream of photons. At its most basic level, the electric force is the continuous exchange of a countless number of photons between charged particles, one photon at a time.

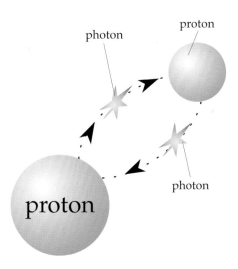

Figure 4 The subatomic 'give and go.'

Based on this exact analogy, a theory for the strong force was proposed in 1935 by Hideki Yukawa (Japan, 1907–1981). Yukawa reasoned that for the exchange of these new 'quanta' to be able to account for the known properties of the strong force, the new entities would have to come in three varieties: electrically neutral, positively charged, or negatively charged. In addition, the electric charge had to be equal in magnitude to that of the proton, and all the entities had to have a mass of about 200 times that of the electron. The need for the three different charges stems from the fact that the nucleons come in two varieties, the proton and the neutron. The 'basketball' can transfer charge from one nucleon to the other or leave the respective charges undisturbed. After some false sightings, the telltale signs of just such particles — coming in three varieties of charges and weighing some 280 times more than the electron — were confirmed in 1947. First named mesons — for having mass between that of the 'heavy' nucleons and the 'light' leptons — they are today called pions, denoted by the Greek letter π, the three charged varieties called, for short, π plus, π minus and π zero. With the confirmation of the pions, the number of particles in the world of the strong force now jumped from two to five — two nucleons and three pions. This was only the beginning. A great proliferation was just getting underway.

Mesons, Baryons and Hadrons

The sighting of pions — that is, the telltale tracks left by pions in the stacks of specially-prepared photographic plates flown high to the edge of the atmosphere, attached to high-altitude balloons — was one of the last hurrahs of the so-called cosmic-ray physics. Cosmic rays refer to streams of ultra-high-energy protons that traverse outer space. Although similar in name to such other 'rays' as gamma rays or X-rays, cosmic rays are not a part of the electromagnetic spectrum that the latter two are. As these protons smash into the nuclei of air molecules, violent collisions among the nucleons involved produces cascades of particles, some old and others new. For a good part of the first half of the 20th century, these naturally-occurring collisions were the only source available to study and discover many new particles. Nothing man-made could match the scale of the energies involved. All detection experiments were done in a non-controlled passive mode: high-altitude balloons loaded with photographic plates were sent up and exposed to the debris of collisions among particles, the plates were brought back down to the ground, and the tracks etched on the plates by particles were examined. Positrons were discovered this way, and pions also. Beginning in the 1950s, the face of the elementary particle changed drastically, from the natural environment of cosmic rays to the man-made, and hence much more controlled laboratory setting, as the new breed of powerful machines called particle accelerators came into being.

Particle accelerators had come a long way by the 1950s, since the days of the hand-held contraptions back in the 1930s. The scale of design and energy output grew steadily, and in keeping with their growth, the names also went through several stages — cyclotron, synchrocyclotron, synchrotron and today's huge colliders. The colliding accelerator at the Fermi National Accelerator Laboratory, called *Tevatron* — located about 50 miles due west of Chicago — is built around circular tracks about 4 miles in circumference. Huge in scale, very expensive to build and maintain, and high-tech in terms of their engineering, the particle accelerators are based, however, on relatively simple low-tech principles. With the help of magnetic force generated by thousands of electromagnets, a beam of charged particles, mostly protons, are kept in circular orbits; protons just go around and around. Along the circular path are sources of electric forces that whack the passing protons to ever-faster speeds. The energy of the protons increases in proportion to their speeds. The accelerated protons routinely reach up to 99.9999995% of the speed of light. That's some acceleration!

Typically, you accelerate one beam of protons in a clockwise circular path and another beam of protons in a counterclockwise path. When all is ready, you bring them smashing head-on; one proton, flying at near the speed of light, rams into another one coming the opposite way, also at near the speed of light. They smash into each other and both 'vaporize' instantly into a ball of formless energy ($E = mc^2$ turns their masses into energy, adding to the enormous energy they acquired from their speed); the formless energy in turn creates, in a dance of energy condensing into mass, all manner of new and hitherto unseen particles. No sooner than created, these new particles subsequently also disappear into energy, turning into a horde of still newer and smaller particles. This process of the mass converting into energy and energy materializing back into lighter masses continues until at the end we are left with only our old friends, the nucleons and pions.

We can try to put this in terms of more familiar things, although the analogy does risk sounding a bit silly. Instead of protons, let's say, yes, we have two avocados, just run-of-the-mill nondescript avocados from the vegetable stand of any grocery store! Suppose we accelerate them to some super-fantastic speed and in one glorious moment bring two of them smashing into each other! They totally demolish each other and in a blinding flash disappear into a puff of formless energy. Just as suddenly, then, the ball of energy materializes back into one big watermelon and two honeydews! In the very next instant, the watermelon and honeydews do their thing; they too disappear in a puff of smoke, only to be reincarnated, in the very next moment, into some grapefruits and oranges. Yes, the downward process continues, $E = mc^2$ doing its thing every step of the way, from grapefruits and oranges to mangos and avocados, with maybe even some scattered plums! In the course of this powerful collision between avocados, we did see, albeit only for brief moments, a string of fruits — a watermelon, honeydews, grapefruits, oranges, mangos, and so on. Most of them did not stay as they were, but did exist for one time in this sequence of creation, conversion to energy and return to matter. Fascinated and totally enthralled, you smash more and more avocados, at faster speeds and thus higher energies, and create all manner of fruits under the Sun, if only for a brief slice of time, hundreds upon hundreds of them!

In this manner, through the late 1940s, 1950s and until the 1960s, a greater number of new short-living heavier particles — heavier than nucleons and pions — have been detected, first by the tens and later by the hundreds. A whole new industry of physics sprang up and went on for decades, alternatively referred to as elementary particle physics or high-energy physics. First we artificially created them in particle accelerators, and then determined all their properties as best we could, carefully cataloging, classifying and placing them into different groups of similar particles. It was like discovering hundreds of new species of plants and carefully cataloging and classifying them into families — a study in botany.

As the new crop of artificially-produced particles transmuted themselves — 'decayed' as they call it in the trade — down the chain to end up as the more familiar nucleons and pions, it soon became clear that among the myriad of these new particles there are two distinct groups: those that are related to nucleons and eventually 'decay' down to them on the one hand and those, on the other hand, related in a similar vein to pions. Each group came to number in the hundreds, and while they were all SIPs, their division into two distinct groups became clearer as more data poured in.

The group that was anchored at nucleons, that is, those that eventually 'decayed' down to nucleons, needed a new name to designate what is essentially the extended family of the nucleons. The name chosen was 'baryon,' from the Greek word 'bary' meaning heavy, with nucleons as the lightest baryons. The other group related to pions were denoted by the name originally given to pions, the 'mesons,' with pions being the lightest mesons. The original family of SIPs consisting of nucleons and pions was now a large extended group containing hundreds of baryons and mesons, and presently the name SIPs was replaced by a new one, the 'hadrons,' meaning strong. As the extended family of the nucleons and pions, the large group of baryons and mesons together came to be called the hadrons. Hadrons, and hadrons only, are the carriers of the strong force; leptons have never heard of it. As far as the weak force is concerned, on the other hand, both hadrons and leptons are affected by it equally.

The extension of the nucleon to the baryon involved more than just a proliferation in numbers and a change in name. The nucleonic charge and its zero-sum rule — the total number of nucleons before and after a reaction to be the same — were extended to all baryons; a new charge called the baryonic charge was invoked for all baryons, including nucleons. All baryons, including nucleons, were assigned the baryonic charge +1, and −1 for all antibaryons including antinucleons; in any reactions the zero-sum rule for the baryonic charge was observed to be strictly upheld. To this date, no violation of the three zero-sum rules — those for the electric, leptonic and baryonic charges — has ever been observed. With the extension of the nucleonic charge to the baryonic charge, and the nucleon-pion system of the SIPs to a hundred or so hadrons, the neat picture of a pair of nucleons had completely evaporated. One was now faced with the headache of having to sort out and make some sense out of the seemingly chaotic world of hadrons.

The Great 'Reverse Engineering' of Hadrons

One way to best understand what transpired next in the development of the physics of hadrons is to make a hypothetical study of how the physics of atomic structure might have developed in a sequence that is exactly the reverse of what actually happened; that is, to run the film backward in a sort of 'reverse engineering.'

Before running the film backward, however, let us first briefly recap on the essential steps by which we acquired our knowledge of atomic structure. First, we discovered the constituents of atoms — the electrons and atomic nuclei — that were held together by a force, an electric force between the charges of the electrons and its nucleus. Then we came to realize that the force was transmitted by its own discrete quanta — photons in the case of the electric force — and it was the continuous exchange of these quanta that represented what we called force. Then there was more. The electrons would gulp down a photon — absorb — and enable themselves to jump up to a higher orbit, and after a while spit the photon back out — emission — and drop back down to their natural habitat, the lowest orbit. An atom would in this way go into one of its 'excited' states with higher energy, and after a while drop back down to its normal 'ground' state, the natural state with the lowest energy. There are many — tens and hundreds — such higher orbits available for the atomic electrons to jump up to, and as they jump up the atom can exist in as many excited states as there are higher orbits filled with electrons. We all 'know' that these excited states are not 'new' atoms, just higher energy forms of the same anchoring atom. We know this as we know the underlying structure of atoms — the identity and properties of the constituents, the force that holds them together, the quanta of the force, and how these quanta were absorbed and emitted as the atom went up and down the 'rung' of the ladder of the excited states.

Now, let's run this film backwards. Without any inkling or sign of possible breakdown in their 'elementaryness,' suppose we kept discovering, at the turn of the 20th century, first tens and then hundreds upon hundreds of 'new' atoms. Things would have been as exciting as they were chaotic; as we began to sift through and sort out patterns of similarity and association, the periodic table of elements would run to tens of pages, containing charts of what we thought were recurring patterns in their chemical properties. After the data were all placed in some recognizable pattern, it might have occurred to someone —bless his soul —that perhaps, just perhaps, all these atoms could not be truly elementary. Might there not be an underlying structure among all this? What if the atoms had internal structure of all their own? If so, what might this be like? You might then try and guess the ingredients (electrons and nuclei), determine what the force between those constituents ought to be (the electric force), develop a grand theory in which the force is represented by its own form of discrete quanta (photons) and how these quanta could be absorbed and emitted by the constituents. And, behold, suddenly things would begin to fall into place. All these atoms were not all distinct and elementary atoms after all; there were a set of anchoring atoms, and all others were simply excited states of these same old anchoring atoms. Out of chaos, we would have brought order, simplicity and a grand new theory of atoms in which a set of new subatomic constituents formed atoms following all the rules of the new atomic physics. It would have been a great triumph, no less of one than the actual history of it!

Well, by now, I am sure you guessed why I ran the film backward. Yes, that is exactly how it happened with the hadrons, in reverse order. After some two decades of high-energy physics, the age of discovering, cataloging and classifying hundreds upon hundreds of 'new' hadrons, it began to dawn on us that perhaps, just perhaps, all these hadrons — including the original protons and neutrons — were not as elementary as we once so firmly believed. We realized that we were instead looking at a horde of 'excited' hadrons, that there were underlying structures that made up all these hadrons, and that, in fact, we were at the threshold of another entirely new layer of matter, just beneath the façade of hadrons.

If so, there would have to be new matter, that which made up not only protons and neutrons but all the new hadrons that we have uncovered in laboratories. The new constituents had to be bound by yet another new force, operating at a much deeper, and smaller, environment — not within atomic nuclei, but in fact within the confines of hadrons; the baryons and mesons. The new force would have to have its own form of quanta, and these quanta would be absorbed and emitted by the new constituents. There would be, yes, a whole slew of 'excited states' of hadrons, and these were the ones that we kept uncovering as 'new' hadrons all this time. It was indeed a grand 'reverse engineering' and that is exactly how, coming into the early 1960s, things unfolded. The new 'underlayer' of matter, the constituents of hadrons, were named 'quarks' (do you have any problem with that?!), the new force between the quarks was called the 'color' force, and the quanta of the new color force — the 'photon of the color force' — would come to be called the 'gluon' (glue holds things together, doesn't it?).

10

The Quark: The Queen of Fractions

The 'grand reverse engineering' of hadrons — trying to deduce the possible internal structure and its dynamics from the observed patterns of hundreds of newly-uncovered hadrons — came to a head in 1964. In that year, often dubbed 'the year of quarks' in the annals of particle physics, a workable new scheme for the internal structure of hadrons was boldly put on the table. At first, the proposal was met with unmasked scorn; the idea was far-fetched and did not have a shred of direct factual supporting evidence; some aspects of it were out-and-out absurd; and the proposed name 'quark' was a joke. Over the next three decades, however, the mountain of evidence accumulated in support of the new theory — albeit indirect and circumstantial — has become so compelling that it has become the accepted standard for the theory of the origin of matter. This then is the story of quarks; how they came to be, why they are — and still remain — so weird, and how the strong and the weak forces came to be reformulated in terms of them.

The Year of Quarks

The proposal that hadrons may be composite structures made up of yet smaller substructures — thereby declaring protons and neutrons to not be elementary at all — was put forward at about the same time in 1964 by two different physicists, both working totally independently of each other. One of the two theories was proposed by one of the most respected and established names in the business, Murray Gell-Mann (USA, b. 1929) of Caltech. Gell-Mann's published paper has become one of the most referred papers of this century, the genesis of a new theory for the structure of hadrons. The new hypothetical substructures of hadrons were christened the 'quarks' by Gell-Mann, and in 1969 he was awarded the Nobel Prize for Physics for his contribution to the systematics of the elementary particles. The name 'quark,' picked by Gell-Mann out of a passage in a novel where it was used as a meaningless cheer, became inalienable in physics nomenclature. The other of the two proposals has a story as sad as any Greek tragedy. It was put forward by an unknown young postdoctoral research fellow, George Zweig (USA, b. 1936). Zweig's paper never got published and later on became one of the most celebrated unpublished works in the history of modern physics; the name he coined for the new constituents — 'aces' — becoming a distant memory. Zweig eventually switched from physics to biophysics.

According the Gell-Mann-Zweig quark model — as the new scheme for the underlying structure of hadrons came to be called— 'reverse engineering' can be very effectively achieved by zeroing in on the following scheme: All known baryons related to and including the original nucleons are considered as composite structures consisting of three quarks (and all known antibaryons related to and including the original antinucleons are considered, as a matter of course, as composite structures consisting of three antiquarks) and all known mesons related to and including the original pions are considered as composite structures consisting of one quark and one antiquark. Furthermore, in order to account for all baryons related to nucleons and mesons related to pions, it was necessary to invoke two different types of quarks. Just as there are two types of nucleons — the proton and the neutron — there must be two types of quarks. What to call the two types of quarks? Gell-Mann did it again and promptly named them 'up' quarks and 'down' quarks.

To begin with, the proposed scheme generated four different family trees each for the baryons and mesons at hand, the extended family of nucleons and pions. For the baryons we had the four combinations, up-up-up, up-up-down, up-down-down and down-down-down, and for the mesons (get a grip of yourself) the up-antiup, up-antidown, down-antiup and down-antidown combinations. Each of these combinations entailed, through their internal dynamics, many excited, higher-energy (and hence higher mass) baryons and mesons of the same combination, and the hundreds upon hundreds of the observed baryons and mesons — related to nucleons and pions respectively— now corresponded to the excited manifestations of the quark-quark-quark and quark-antiquark combinations, respectively. That, in essence, was the simplicity and power of the Gell-Mann-Zweig quark model, the climax of the grand 'reverse engineering.'

One of the first things that grabs one's attention is the intrinsically unstable nature of the mesons, when viewed as composite structures made up of one quark and one antiquark. We know from the very definitions of the matter-antimatter dichotomy that they greet each other with a violent and self-vaporizing disappearing act known as the matter and antimatter annihilation. If a quark should be confined within a small volume of space, the interior of a pion, with its matching antiquark, well, it wouldn't take a rocket scientist to figure out that in no time they would annihilate each other and in a puff of blinding light the pion would be no more. Well, that is actually just as well, because all mesons — including the pions themselves — are known to have extremely short natural lives. These particles 'die' almost as soon as they are 'born': once created in a high-energy collision among particles in the particle accelerator, the longest-living meson, a pion, lasts no more than about 30 billionths of a second, 30 nanoseconds in today's parlance. As we discussed previously, pions are 'the photons for the strong force'; the continual exchange of pions being the mechanism responsible for the strong force between nucleons. Despite such an incredibly short lifespan, one need not worry about a pion 'dying' between two nucleons; 30 nanoseconds are more than 'long' enough to cover the short internucleonic distances involved.

The Queen of Fractions

What truly set these quarks apart from all other particles that had hitherto been known was not so much their names — which are admittedly a bit too playful, but rather the properties that they had to be endowed with to do the job they were invoked to do. A closer examination will reveal that their destined properties are truly out-and-out weird; let us now examine those.

First, the matter of the baryonic charge. Let us recall that the baryonic charge of +1 is assigned to all baryons, and −1 to all antibaryons, in a straightforward extension of the nucleonic charge assignment of +1 for nucleons and −1 for antinucleons; the baryonic charge subsumes the nucleonic charge and is an immediate extension of the latter. Nucleonic charge, in turn, was assigned to nucleons on the strength of a zero-sum rule that was strictly adhered to by any reaction involving nucleons — that the total number of nucleons before and after a reaction should and do remain the same. By the same token, the zero-sum rule for the baryonic charges stipulates that the total number of baryons before and after any reaction be the same. To this date, no violation of this rule of the strict conservation of the baryonic charge has ever been observed, and it serves as one of three zero-sum rules for the charges — the electric, baryonic and leptonic — that, for reasons not wholly understood, form the bedrock for the zero-sum rules for the subnuclear world.

According to the Gell-Mann-Zweig quark model, baryons are to be made up of three quarks, and this dictates the baryonic charge of quarks to be exactly one-third, yes, 1/3 of a baryonic charge. This is one of the reasons that earns quarks the label 'fractionally-charged.' Of course, this is more a matter of convention than anything else; with 20-20 hindsight we could have re-defined the baryonic charge of a baryon, say, a proton, to be +3. Then the baryonic charge for quarks would be more 'reasonable' at +1! It is all a matter of relative scales and there is nothing fundamental about either choice. The situation is in fact reminiscent of the case of the electron spin versus the photon spin, the former being one half of the latter. We could have defined the spin of a photon to be 2 and the electron spin would have been 1; as it is, we have defined them to be 1 for the latter and 1/2 for the former. Strictly speaking, then, the spin of the electron is also 'fractional,' for that matter. The use of the label 'fractional,' is, however, reserved — at least in the physics of elementary particles — for the fractions involving one-thirds, a property exclusive to quarks.

The fractional character of quarks becomes even more weird when it comes to the question of their electric charge assignments. Unlike either baryonic or leptonic charges, which have no large-scale applications in our daily world, the world of units related to the unit of electric charges — everything from volts and amperes to ohms and watts — has been solidly established, and in this scale the electric charges for the proton, neutron and electron are +1, 0, and −1, and that's that. It would be very unwise to have it otherwise. Now, of the four possible combinations of the three-quark configurations — uuu, uud, udd and ddd, using 'u' and 'd' for the up and down quarks — the Gell-Mann-Zweig model assigns the uud combination to the proton and the udd combination to the neutron. A simple calculation (the electric charges of two up quarks and one down quark should add up to +1, and, likewise, those for one up quark and two down quarks should add up to zero, for the neutron) will show you that the electric charge assignments for the up quark must be +2/3 and that for the down quark −1/3. Yes, you read that right, it is positive two-thirds for the former and negative one-third for the latter.

Note that the difference of one whole unit, as the difference between +1 and 0, is strictly maintained — between +2/3 and −1/3. The scale is shifted down 1/3, from +1 to +2/3 and from 0 to −1/3. Other than that it is no big deal, except for one thing: There has never been anything like this before, in the history of particle physics, even in our wildest dreams! There is no denying that quarks have earned the title 'fractionally-charged particles'! The baryonic and electric charges for quarks, compared to those for nucleons, are then as follows:

	Baryonic charge	Electric charge
Quarks		
Up	+1/3	+2/3
Down	+1/3	−1/3
Nucleons		
Proton	+1	+1
Neutron	+1	0

In addition to the proton (up-up-down) and neutron (up-down-down), we were in need of nucleon-like particles corresponding to up-up-up and down-down-down compositions with the electric charges +2 units and −1 unit, respectively. Sure enough, these were soon uncovered, again among the debris of high-energy collisions among nucleons, thus strengthening the case for the Gell-Mann-Zweig quark model. They were named 'delta-double plus' and 'delta-minus.' The four combinations for mesons, on the other hand, are the positive (up-antidown), negative (down-antiup) and two neutral combinations (up-antiup and down-antidown). They correspond to the positively-charged pion, the negatively-charged pion, the neutral pion and the second neutral meson, discovered soon thereafter and named 'eta.'

Quarks, Show Thyselves

Unlike the baryonic charge — or, for that matter, the leptonic charge as well, which is somewhat arbitrarily assigned on the strength of its zero-sum rule — the electric charge is something that can be easily and readily measured. There are all manner of ways and devices that enable us to measure, directly and very precisely, any amount of electric charge of a particle or an ionized atom, however miniscule it may be. A charged particle moving through an electric field, a magnetic field or some suitably configured medium, be it a gas, liquid or solid, leaves behind enough clues for us to make very precise measurements of its electric charge. Differentiating a particle with the electric charge +1 from any other particle whose electric charge is +2/3 or −1/3 is a task physicists find almost as easy as downing a Coke and a cheeseburger.

If there are particles with their electric charges so different from all other known species, they should have been discovered with relative ease. Guess what? Since their original introduction in 1964 till this day, not a single trace of any evidence has been recorded for the direct observation and measurement of the type of fractional charges that are associated with quarks. Not one. And it was not due to lack of trying (a Nobel Prize is practically reserved for the first person to nail it down and have it confirmed!). Ever since 1964, countless efforts have been made to detect the unique trace of quarks in just about all possible potential environments: in the aftermath of a high-energy collision of protons, antiprotons, electrons and positrons — the gold mine where so many new particles have been discovered; inside every high-energy accelerator in the world; on top of mountains; at the bottom of deep seas; in the depth of the Antarctic ice; and far in outer space. Scientists have not only examined samples of moon rocks, brought back by the Apollo astronauts, but also the innards of oysters scooped up from some of the deepest ocean beds. What's with the oysters? Oysters are known to eat just about anything they can. So why not some quarks as well? To this very day, despite all-out efforts made at laboratories worldwide, no trace of objects that even remotely resemble the one-thirds fractional electric charges have ever been observed. Not a single one.

What about frequent reports that this or that type of quark has been discovered then? They are all indirect 'discoveries': when physicists discover a new meson that does not fit their expectations of a composite system made of the known quarks and their antiquarks, it is claimed as the discovery of a composite system of an entirely new species of quark and its matching antiquark, and the properties of the new quark are then inferred from the observed properties of the new meson. You infer the existence of new constituents by discovering new compound systems; to put this in terms of atoms and molecules, it is like inferring the existence of a new atom by discovering a new molecule not seen before and that does not fit the expectations of a compound system of any other known atoms. All so-called 'discoveries' of quarks to date have been indirect inferences in this manner, however compelling the inference might be.

If quarks are not to be 'seen' — that is, detected and measured in their unattached and isolated single form, but seen only in threes, as in a baryon — a natural question might then be, what about a system of two — not three — quarks. If three quarks can stick together to form a baryon, it would seem very natural to look for particles made up of two quarks sticking together, up-up, up-down and down-down combinations. Remembering the electric charges of up and down quarks being +2/3 and −1/3, respectively, these two-quark systems would have to have their charges +4/3, +1/3 and −2/3, respectively. As far as the non-traditional signature of their electric charges are concerned, these two-quark combinations would stand out just as much from the ordinarily-charged particles as a single quark. Both single quarks and two-quark systems are endowed with this 'fractionality' by the thirds. To add insult to injury, nothing of this kind has ever been detected either. Zilch. Threes, yes. Ones and twos, no.

The Triumph of Shadows

The total absence of the direct observation of a single quark — of any kind and by its unattached self — would have been seen, by any traditional standard, as a fatal flaw in the theory. It is a little difficult to talk about real things — protons and neutrons inside atomic nuclei — that are made up of three little shadows — quarks — that one can never confirm in a laboratory setting. Something like this has no equal in the annals of physics. Molecules can be broken up into their constituent atoms with relative ease by chemical reactions; atoms can be just as easily broken up by ripping out the electrons — referred to as ionization — from atoms, leaving behind bare nuclei. The physical characteristics — mass, electric charge, and others — of each individual constituent — electrons and nuclei in this case — are precisely measured. Going downscale further, an atomic nucleus can be split apart — by hitting it hard by another nucleus for example — into its constituent nucleons. When a deuterium (the heavy hydrogen that forms heavy water when combined with oxygen) nucleus that consists of one proton and one neutron is impacted by a proton — the hydrogen nucleus — it readily splits into a pair consisting of an unattached proton and a neutron. The characteristics of protons and neutrons are determined, recorded and confirmed by all manner of experimental measuring devices. This is what is meant by a direct observation. And now we come to what appears to be the final layer of matter — quarks inside protons and neutrons — and we are faced with the awkward prospect that we may never be able to measure directly all the physical properties of the final frontier of matter, the quarks! It is a strange turn of events indeed.

Despite such a profound flaw, the theory of matter based on the quarks as the final ingredients of matter is now virtually a completely accepted standard. How did that come about? The answer to this question is simple: over the three decades since its initial inception in 1964, the amount of data accumulated that can be effectively explained in terms of the quark constituency is unassailably impressive. The number of successful explanations and predictions in terms of the theory of quark constituency has proved to be too compelling not to accept. With some still unanswered loose ends here and there, virtually all aspects of particle physics — baryons as systems of three quarks and mesons as a quark and an antiquark — have been successfully explained by assuming quarks to be real. This is referred to as the success of quarks in explaining the baryon and meson spectroscopy; the spectroscopy borrowed from atomic spectroscopy, the latter successfully explained by the electron-nucleus constituency of atoms.

In addition to this spectroscopic success, more 'direct' indirect evidence in favor of quarks has been obtained by, of all things, repeating what Rutherford did to discover the atomic nucleus in the first place back in 1911. In a remarkably similar setting — on a much smaller scale, of course — a team of physicists bombarded protons and neutrons with very high-energy beams of penetrating electrons. These electrons were able to penetrate the innards of a proton or a neutron, and just as Rutherford discovered the nucleus, the explorers of the 1970s were able to again map out the existence of highly-concentrated points of electric charge lurking within the innards of a nucleon. They were not able, no matter how hard they tried, to pry loose single quarks from within the nucleon, however. Touted generally as the 'discovery' of quarks, the work earned three physicists — Jerome Friedman (USA, b. 1930), Henry Kendall (USA, b. 1926) and Richard Taylor (Canada, b. 1929) — the Nobel Prize for Physics in 1990. The prize citation stated, correctly, only that their work "has been of importance for the quarks model of particle physics."

Since the discovery of the electron in 1897, we have come a long way. First the structure of the atom in terms of the nucleus and electrons; then the structure of the nucleus in terms of protons and neutrons; and finally the structure of nucleons in terms of quarks — the ups and downs of them. The irony of the final layer, if it is indeed the final one, is not lost, however; quarks are not seen, that is, they have never directly been observed. But quarks it is, and our relentless pursuit for the origin of matter now rests on our understanding of quarks and leptons as the most basic constituents of matter, and of the two new nuclear forces, the strong and the weak, in terms of them.

11

The Origin of Quarks and Gluons

The view that nucleons — as well as pions and, by extension, baryons and mesons — are not really elementary but rather composite structures of quarks was just as epoch-making as the discovery that atomic nuclei are made of protons and neutrons. The first logical step was then to re-interpret the nature of the two forces, the strong and the weak forces, that reside deep inside nuclei. Of the two, the weak force is all-inclusive, in that it affects all particles — hadrons and leptons alike. The strong force, on the other hand, is exclusive; only the hadrons are members of the 'strong' club; leptons are unaffected by it. Whatever attributes of quarks are responsible for the weak force must be present also in leptons, but, on the other hand, the attribute of quarks responsible for the strong force must be something that is exclusive to quarks.

The Quark Origin of the Strong Force

The strong force between, say, two protons — treated hitherto as a fundamental force between two point-like and elementary protons — becomes, in terms of the quark constituents, a rather complicated affair. As two protons come close to each other (close enough for the strong force to come into play) the picture becomes a three-on-three situation. A trio of quarks comprising one proton —up, up and down —meet up with another trio of up, up and down, which make up the second proton. Whatever the nature of the new forces between quarks is, it is now a three-on-three play that represents a set of nine pairs of interquark forces, the proton-proton strong force corresponding to an average of four pairs of up-up force, four pairs of up-down force and a pair of down-down force. Things are getting a bit complicated.

The first thing that presents itself is the conclusion that whatever the nature of the interquark force is, it cannot be the same as the strong force between hadrons. The strong force is strong all right, but clearly not as strong as our ability to crack open nuclei; armed with high-energy particle accelerators that can propel protons to near the speed of light, we can crack open atomic nuclei any day of a week, 24 hours a day! Just like a walnut being smashed open with a sledgehammer, atomic nuclei split open and spew out their constituent protons and neutrons in all directions. Quite to the contrary, and much to our disappointment, we have not been able to repeat this feat with quarks; from the early 1960s until now, no one has succeeded in cracking open a proton or a neutron to release, capture and examine a single solitary quark. This is living proof that the interquark force is not only new but quite different from the strong force we already know.

The Color Charge of Quarks

A new force — the interquark force that is the precursor of the strong force — makes it necessary for us to invoke a new 'charge,' a new kind of charge that is exclusive to quarks alone. It must clearly be neither the electric charge (the new force is not in any way related to good old electric force), nor the baryonic charge (the new force is not the strong force whose strength is reduced by one-third), nor the leptonic charge (the new force has nothing to do with leptons). The first hint that the world of quarks may involve attributes outside of any type of charge then known came in 1965 — a year after the year of quarks — when it was proposed by Moo-Young Han (Korea, USA, b. 1934; the author of this book) and Yoichiro Nambu (Japan, USA, b. 1921) that the quarks harbored a hitherto unknown tri-valued attribute. Quite unlike any of the previously-known charges — which are basically two-valued attributes (positive and negative) — the new tri-valued attribute of quarks was something entirely new, a set of three different values adding up to zero.

If you represent the two previous kinds of charge — either electric or baryonic — as points on a line, positive values to the right of the origin and negative values to the left, the pair of positive one and negative one, say, sum up to zero at the center of the line. It is in this sense that they can be referred to as being 'one-dimensional' charges; positive and negative numerical values represented on a line. The new tri-valued attribute of quarks, proposed by Han and Nambu, is the first known extension of the concept of a charge from one- to two-dimensional values. Imagine three vertices of, say, an equilateral triangle. With respect to the center of the triangle, the vertices have three different values that sum up to zero, the center of the triangle. The designation of the three 'values' — three vertices — requires more than simple one-dimensional positive and negative numbers. Likewise, it would take more than a set of positive or negative numerical values to represent these new attributes.

It took several years for this idea of a new tri-valued attribute of quarks to gradually evolve into the realization that we were indeed talking about the emergence of a totally new kind of charge that was an exclusive property of quarks and quarks alone. It was in 1972 that this concept was elevated to the full-fledged status of a new and independent charge; a charge with three non-numerical 'values' that, however, must add up to zero for nucleons and pions, and hence for all hadrons, baryons and mesons. In other words, in the world of ordinary particles which can be detected and observed, no trace of this new quark charge is left; it is something that operates in the world of quarks alone. The name for this charge was coined by Gell-Mann — yet again — as the 'color' charge. The three different 'values' of the color charge were christened 'red,' 'green' and 'blue.' No numerical value could be assigned since it is a 'triangular' attribute; the choice of the name 'color' appears at first sight to be rather whimsical, having nothing whatsoever to do with the meaning of the word 'color' in its normal usage. The name does make sense, however, in its parallel with the three primary colors — an equal mixture of the three primary colors turns into colorless white light. The red, green and blue color charges sum up in equal proportions to be colorless, color charge neutral — like three vertices of a triangle collapsing onto its center.

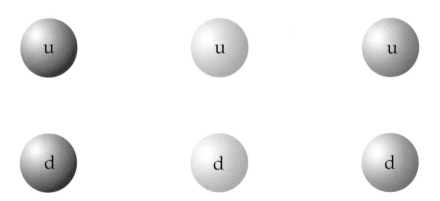

Figure 5 The primary colors. Each quark — here the up and the down quarks — comes in three different 'color charges': red charge, green charge, and blue charge.

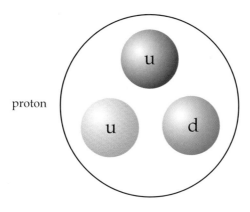

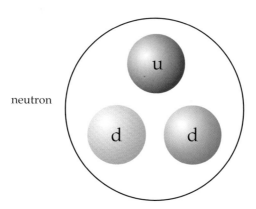

Figure 6 The 'colorful' quark picture of nucleons. The red-green-blue color charge distribution, as shown, is just one of three possible distributions, the other two being the green-blue-red and the blue-red-green options. A nucleon is in fact the composite of all three distributions.

The table for quarks and nucleons therefore becomes like this:

Quarks	Baryonic charge	Electric charge	Color charge
Up	+1/3	+2/3	red, green and blue
Down	+1/3	−1/3	red, green and blue
Nucleons			
Proton	+1	+1	none
Neutron	+1	0	none

The Color Force and Gluons

The force among the newly-christened 'color' charge — a tri-valued attribute — is what holds quarks together inside hadrons — nucleons in particular. Considering that we have not been able to dislodge a single solitary quark out of a nucleon, we guess that the attraction among quarks must increase with the distance between them; that is, this force — that came to be called the color force or chromoforce — becomes stronger the more you try to pull quarks apart. It would take an infinite amount of force to free one quark completely from the other two quarks inside a three-quark nucleon. At least, that is the current 'party line,' and it will stay valid until and if we can ever pry open a nucleon and actually dislodge a single quark out of it.

The subatomic understanding of the electric force, as we described in previous chapters, is that the force is the manifestation of a continuous exchange of photons between the charges involved. It was this concept that led to the successful prediction, and subsequent confirmation, of the existence of pions, the continuous exchange of which between the baryonic charges of nucleons was taken to be the mechanism underlying the strong force. Now, the strong force among hadrons is cast as the collective behavior of a host of the color forces among the color charges of quarks, and in an exactly parallel manner the new force necessitates the introduction of particles whose continuous exchange among the color charges corresponds to what it is. The set of these new particles, the 'photons' of the color force, if you like, are called the gluons. The name comes, again, courtesy of none other than Murray Gell-Mann! Shuttling between sets of three colors— red, green and blue— the gluons all carry color charges themselves, and there has to be not one, not three, but eight of them. Of the nine possible color charge pairs, one that corresponds to the equal mixture of the three colors— the colorless white combination of the three primary 'colors'— must be subtracted out, and this leaves eight different combinations of three colors. The set of eight 'photons' of the color force is hence referred to as the 'octet' of gluons.

As far as the strong force of hadrons is concerned, this is where we are at. At the bottom of everything, there are the quarks which, among other charges— electric as well as baryonic— carry a set of three color charges, exclusive to quarks alone, and a color force acts among these color charges. The color force manages to collect quarks only in certain combinations — three quarks for baryons, quark-antiquark pairs for mesons, and three antiquarks for antibaryons — to form the heavy particles. Protons and neutrons, in particular, attract each other to form atomic nuclei, with so many protons and so many neutrons. And the collective manifestation of the color forces among the constituent quarks corresponds to what we call the strong force. That, according to our latest understanding, is the story of the once-fundamental strong force.

What About the Weak Force?

Would the same thing that happened to the strong force also happen to the weak force? A parallel would have been something like this: an entirely new class of force must be assumed among the quarks — in addition to the color force — such that the net result of the aggregate of this force corresponds to the observed transmutation of a neutron into a proton, that is, the work of the weak force. In that case it might be equally compelling to invoke yet another new charge for the quarks that would be responsible for this new force, just as the color charge is the source of the color force.

Curiously the extension of the weak force into the realm of quarks took a path that is entirely different from that of the strong force. The extension is as simple as it is direct: the transmutation of a neutron into a proton with the accompanying emission of a pair of leptons — an electron and an antineutrino — is 'explained' as the result of a down quark inside a neutron transmuting itself into an up quark. As a down quark turns into an up quark, the electric charge changes from $-1/3$ to $+2/3$, the gain of one unit of the electric charge getting balanced out by the one unit of negative charge of the emitted electron. The weak force that turns a neutron into a proton, with the creation and subsequent emission of a pair of leptons is, according to this understanding in terms of quarks, just the manifestation — at the nucleon level — of one down quark turning into one up quark inside a neutron. The creation and subsequent emission of the electron-antineutrino pair taking place at the quark level within a neutron.

The explanation of the weak force has thus gone through two stages of reduction in scales; first, the transmutation between atomic nuclei was explained in terms of that between a neutron and a proton within the nucleus, and now one more step downscale to that between a down and an up quark within the neutron. The rules remain the same, only the identity of the participants has changed. Some would say that the rules of the game have not changed, but that the goalposts have been moved further and further back!

The basic picture of a force as a continuous exchange of streams of a special class of particle — the particles of force — should apply to the weak force as well. In an analogy with the photons for the electric force and the newly-christened gluons for the color force (or the erstwhile picture of the exchange of pions among nucleons), the particles of the weak force were introduced as early as the 1950s. Originally only two charged types were invoked, called W-plus and W-minus, but soon a neutral type was also needed and was formally introduced in the 1960s. This was named Z-zero. The threesome — two charged Ws and one neutral Z — were confirmed in the 1980s. According to this picture, the weak force manifests itself like this: a down quark turns into an up quark by emitting a W-minus, again neatly obeying the zero-sum rules for other charges (electric, baryonic as well as color), and the W-minus after only a brief existence turns into an electron and an antineutrino. The zero-sum rules for all the charges are neatly obeyed — electric, baryonic, leptonic and color. The color charge plays no part in the weak dance. The 'anti' side of the process likewise goes like this: an up quark turns into a down quark, emitting a W-plus, which in turn changes into a positron and a neutrino. Everything is cool.

The Origin: Two Pairs, Quarks and Leptons

A century questing toward the origin of matter, from the discovery of the electron in 1897, the shattering of the myth of the indivisibility of atoms, right up to today's knowledge, has brought us to one fundamental picture. All matter, as far as we can ascertain, is in the end based on two pairs of substance; a pair of quarks and a pair of leptons. The pair of quarks, named up and down, is all that is needed to constitute the nucleons, protons and neutrons; the pair of nucleons in turn is all that is needed to make up all the atomic nuclei that exist in nature. Electrons surrounding these nuclei form what we call atoms; atoms interlock to form molecules, molecules interlock with each other to form yet larger molecules. Some remain in the freer form of gas, others coalesce into fluid liquids and still others form hard solids. Every matter in the known expanse of the Universe boils down to its essence — a pair each of quarks and leptons.

At the level of quarks, we have three forces in action: the color force among them mediated by an octet of colorful gluons; the weak force doing the dance of transmutation, emitting electrons and neutrinos, mediated by the Ws and Zs; and the good old electric force between any pair that carries the electric charge, a stream of photons shuttling back and forth in between.

At the level of nucleons, the electric and weak forces remain the same, albeit between nucleons rather than between quarks, but the color force among the quarks is averaged out to be the strong force; the baryons affecting each other mediated by the exchange of mesons, the force between nucleons being its prototype.

At the level of atoms, the traces of the weak and the strong forces disappear within the shields of the nucleus, and the only force that determines the atomic characteristics is now the electric force between the orbiting electrons and the positive charges of the nucleus. From this point on to bulk matter of all sizes and shapes, the electric force rules supreme. The electric force between the outer layers of electrons of constituent atoms is what determines all physical and chemical characteristics of molecules, and by extension all matter. As things become large — rocks, water, rivers and mountains, and the Earth and beyond — the force of gravity asserts itself, eventually on the scale of the Universe, and overrides all other forces.

The three forces — the electric, the weak and the color — correspond to continuous exchanges between participating particles to a set of one, three and eight particles of force, respectively; the photon, the Ws and a Z, and an octet of gluons, respectively. The essence of the electric and color forces are embodied in their respective charges, the electric and color charges each with their own strict zero-sum rules. The underlying nature of the charges for the weak force, however, remains unclear. Whereas the electric and color charges are defined as the source of their respective forces, the fact that the weak force affects all particles — quarks and leptons both — rules out the possibility of either the color or leptonic charges as being the charges of weak force. We have yet to identify such a charge that could be universally carried by both quarks and leptons.

This then is the grand picture of the origin of matter that we have arrived at after a century of quest, and as far as we can ascertain it is correct and represents the latest of our knowledge. Whether it is the 'end of the line,' the ultimate origin of matter of not, we cannot be completely certain, but it is the best that we have been able to come up with as of today — notwithstanding the excruciating puzzle that for all its fundamentality, not a single isolated quark has yet been detected. No doubt, the quest for the origin of matter will continue on well into the next millennium.

Epilogue

More Quarks, More Leptons and More Charges

In this book I have traced the search for the origin of matter, spanning some one hundred years, from the days of electrons and photons to today's quarks and leptons. The belief that at the bottom rung of the ladder of matter, everything is made up out of quarks and leptons represents the latest of our knowledge, the world in which three different forces of nature — the electric, the weak and the color — intermingle with one another under strict zero-sum rules for various charges.

The final picture we end up with involves only one pair of quarks — the up and down quarks — and one pair of leptons — the electron and its associated neutrino. As far as all known matter that exists in the Universe today is concerned, this foursome of two pairs is more than sufficient to account for them — nucleons, atomic nuclei, atoms, molecules, bulk matter, mountains and rivers, planets, galaxies ... everything. This is not to say, however, that this foursome is really the end of the story. There are others (in fact, uncomfortably many others) and the mystery, if anything, widens.

It turned out that as we attained higher collision energy between nucleons, made possible by ever higher-energy particle accelerators, we were able to artificially create and produce a horde of new and hitherto unsuspected particles much heavier than nucleons. And their properties seemed to require explanations in terms of newer species of quarks, above and beyond the ups and downs that make up the nucleons. To be sure, these new and heavier species were really more quasiparticle types, having indescribably brief lifespans — some live billionths of billionths of a second, some others billionths of second, and still others have relatively 'long' lives of millionths of a second. They do not make up any bulk matter that exists in the Universe today, but they serve to illustrate what might have been in the early stages of the Universe — growing, cooling and fast-expanding — and it is in this capacity that they hold clues that will one day help unlock the ultimate truth about the origin of matter.

When we apply the same analyses to these quasiparticles, it soon becomes clear that we need many more new species of quarks, and new species of leptons as well. The list becomes staggering. We need at least four more quarks, named 'strange,' 'charm,' 'top,' and 'bottom.' To go with them, we need four more leptons — the muon, the tauon (also called the heavy lepton, something of an oxymoron since the name lepton comes from the Greek, lepto, meaning light), and two types of neutrinos associated with each, dubbed the muon-type neutrino and the tauon-type neutrino. The two new types of neutrinos necessitate assigning an additional tag to the original neutrino as the electron-type neutrino. The muon had actually been around for a while, and was discovered first in the late 1940s. Only in the 1970s, however, was it given its rightful place in the scheme of things.

As the number of species of quarks and leptons increased from the basic four to twelve, it necessitated — among other things — something of a population explosion in the new charges for all these new members, and the associated invocation of their respective zero-sum rules. First came the 'strangeness charge' for the strange quark, followed soon by the 'charm charge' for the charm quark, and later still the 'top charge' for the top quark and the 'bottom charge' for the bottom quark. The leptons were not to be outdone, and we invoked the muonic charge for muons and muon-type neutrinos and the tauonic charge for tauons and tauon-type neutrinos. Eelementary particle physics became a bazaar of all manner of conserved charges.

Our latest understanding of the whole situation is compactly summarized in what is referred to as the Standard Model of Particle Physics. It is based on what I have just alluded to: six quarks (up, down, strange, charm, top and bottom), six leptons (electron, muon, tauon, and three types of neutrino), and three forces (electric, weak, and color). From the days of electrons and photons to the Standard Model, throughout the history of elementary particle physics, the one guiding principle that has been applied over and over is the principle of conserved charges, the basic attributes of the elementary particles called charges and their respective zero-sum rules. The answer to the physical meaning of all the conserved charges of the elementary particles, along with that of the mass of particles, will hold the key to our understanding the ultimate origin of matter, and, by extension, the origin of the Universe itself.

Appendix 1

Annotated Chronology

The waning years of the 19th century, the period of several years from 1895 to 1900, marked one of the most significant turning points in the history of physics. The period signaled the drawing to a close of what we now refer to as classical physics — the 250 years of developments of classical mechanics and the classical theory of electromagnetism at the hands of such giants as Galileo, Newton, Coulomb, Faraday and Maxwell. In its wake came the first signs of what would come to be called modern physics, the 20th century physics of epoch-making discoveries — relativity, quantum physics, atomic, nuclear and elementary particle physics. From the discovery of the first known subatomic particle — the electron, in 1897 — to today's understanding of the origin of matter in terms of quarks and leptons, the past 100 years of physics represent unprecedented advances in both the scope and depth of our understanding of nature. In this Appendix the most important events and people of this period are chronicled with brief annotations.

1897: Discovery of the electron

While studying the nature of the 'mysterious' glow of gases (neon signs), Joseph John Thomson (England, 1856–1940) discovered in 1897 the particle of electricity that he named the electron. Electrons being smaller and lighter than atoms (considered till then to be the final and indivisible constituents of matter), their discovery implied, for the first time, an inner structure for atoms. The discovery of electrons thus marked the first heralding of atomic physics as we know it today. Thomson was awarded the Nobel Prize for Physics in 1906. The electron, the oldest elementary particle, remains the lightest of all elementary particles, with the exception of the handful that have no mass.

1900: The quantum of light

The energy of radiation, be it the warmth of sunlight or the intensity of a broadcast wave, had been assumed to be continuously valued, rising and falling as smoothly as the mercury column in a thermometer. This was considered self-evident. In a shattering discovery, Max Planck (Germany, 1858–1947) reported in 1900 the unthinkable opposite: when examined in microscopic amounts, radiation energy was not at all smoothly varying but changed in discontinuous spurts; it consisted of distinct individual units. Planck named these units — the specks of radiation energy — the quanta, the discrete quantities of energy. The word 'quantum,' theretofore an obscure word, would never be the same again: physics took it over and virtually monopolized it — quantum physics, quantum jumps, quantum mechanics, quantum fields, and in some cases even the art of quantum meditation! Planck was awarded the Nobel Prize for Physics in 1918.

1905: Relativity

In two papers, published three months apart in the fall of 1905, Albert Einstein (Germany, Switzerland, USA, 1879-1955) unveiled his theory of relativity. The Galilean-Newtonian presumption of separate and independent space and time fell by the wayside. The two were in fact intertwined, and closely influenced each other, drastically altering our view of the Universe. In the second paper, he derived the energy-mass equivalence relation, $E = mc^2$, perhaps the most famous formula of the 20th century. The theory of relativity was so revolutionary that it wasn't even mentioned in the citation when Einstein was awarded the 1922 Nobel Prize for Physics.

1905: The idea of a photon

One of the many profound corollaries of the theory of relativity was that an object could very well have energy and momentum even though it had no mass at all. The theory showed that the idea of a massless particle was just as acceptable as that of one that has mass. Upon this, Einstein took the idea of Planck's quanta one step further and elevated it to massless but bona fide 'particles' of radiation. Einstein renamed quanta the photons, the particles of light— no mass, but particles nevertheless. To this day, one finds 'resistance' to this whenever we see descriptions of photons as the 'particle-like' quanta of electromagnetic radiation— 'particle-like' but not quite 'particles'!

1911: The atom and its nucleus

In May of 1911 Ernest Rutherford (New Zealand, Canada, England, 1871-1937) announced his unlocking of the atomic structure: an atom consisted of a positively-charged central core (the atomic nucleus) and a group of electrons orbiting around it. An atomic nucleus is as much as 100,000 times smaller than the atom that it is the center of, but comprises about 99.98% of its mass. Such concentration of mass strikes an interesting parallel with our own solar system; the Sun accounts for about 99.86% of the entire mass of the solar system. Rutherford, honored by the 1908 Nobel Prize for Chemistry for his work on the disintegration of radioactive substances, never did receive a Nobel Prize for Physics.

1913: The Bohr atom

Niels Bohr (Denmark, 1885–1962) put everything together and came up with the first successful quantitative model for atomic structure. In a grand synthesis of the ideas of Thomson (the electron), Planck (the quantum), Einstein (the photon) and Rutherford (the atomic nucleus), Bohr put together the planetary model for the hydrogen atom, in which a lone electron revolved around the proton in one of the discretely-spaced orbits. As the electron jumped up or down between its orbits — up and down rungs of a ladder — photons are either absorbed or emitted. The quantitative success of this so-called 'Bohr' model of the atom was right on the money. Bohr's work, which earned him the Nobel Prize for Physics in 1922, set in motion the search for a more genuine theory for atoms — quantum mechanics.

1925: The exclusion principle

As Bohr's idea was applied to other atoms, it soon became clear that, in order to explain the recurrence patterns apparent in the periodic table, some new guiding principle was needed. In January of 1925, Wolfgang Pauli (Austria, Switzerland, 1900–1958) proposed just such a principle: no two electrons can be completely identical to each other within one and the same atom, that is, they are either in different orbits or, if in the same orbit, then they must differ from each other by at least one attribute. This exclusivity of electrons — no two can be completely alike within a common atom — helped make sense out of the periodic table of elements. Pauli was awarded the Nobel Prize for Physics for this discovery in 1945. To this day, no one is quite sure of its physical origin, but no deviation from this exclusion principle has ever been observed.

1925: The discovery of the electron spin

In November of 1925, two young Dutch physicists, George Uhlenbeck (Netherlands, USA, 1900-1989) and Sam Goudsmit (Netherlands, USA, 1902-1978), uncovered a major find: electrons possessed a hitherto unsuspected property dubbed 'spin.' A standard metaphor for the spin is the rotation of the Earth about its north-south axis. An electron, as it orbits around a nucleus, also spins about its own axis. We can stretch the metaphor a little bit further; the electron spins one way or the other, clockwise or counterclockwise. The idea of the electron spin provided an interpretation of the exclusion principle. Two electrons in a common orbit around a nucleus manage to remain different from each other by the two spin directions — if one spins clockwise, then the other must necessarily spin counterclockwise, and there is no room for a third one since it will run foul of the exclusion principle with one of the first two. Uhlenbeck and Goudsmit never received Nobel Prizes for their discovery, which was indispensable to our understanding of the atomic structure.

1927: The uncertainty principle

Perhaps the best known of all principles in quantum mechanics, the uncertainty principle, was formulated in 1927 by Werner Heisenberg (Germany, 1901-1976). He theorized that it is inherently impossible to determine both the position and the speed of a particle to a level of absolute accuracy. The more accurately you measure one, the less you know about the other. Paraphrased more broadly, the uncertainty principle is often stretched as: "if you know what it is, then you don't know what it is doing, and, conversely, if you know what it is doing, then you don't know what it is!" This is a bit of an exaggeration, but it captures the essence. As the most dramatic departure from the Newtonian axioms of classical physics, the uncertainty principle is perhaps the second most widely known of 20th century physics, second only of course to the formula $E = mc^2$.

1928: The prediction of antimatter

In 1928 Paul Dirac (England, 1902–1984) found a way to bring quantum mechanics and relativity under one roof. The so-called relativistic quantum mechanics — rules and guidelines for the world of the ultra-high-energy and ultra-fast subatomic particles. There was one catch, however. In order for the new theory to make sense, there had to be antimatter. An anti-electron had to be a 'mirror-image' of an electron, with the same mass but exactly the opposite electric charge — a 'positively-charged' electron. It would be another four years before such a particle was uncovered.

1932: The confirmation of antimatter

In 1932 Carl Anderson (USA, 1905–1991) was able to identify the telltale tracks of, yes, the 'positively-charged' electron, in films exposed to cosmic rays at high altitudes. The 'positive' electron was named the positron, and indeed it was the long-awaited antiparticle of an electron. Anderson was awarded the 1936 Nobel Prize for Physics for the first-ever discovery of antimatter.

1932: The discovery of the neutron

As early as the 1920s it became clear that the masses of constituent protons alone could not account for the entirety of the mass of a given nucleus. Something else had to be inside the atomic nuclei to make up the rest of their mass, and the possible existence of a neutral particle with about the same mass as a proton came to be suspected. In 1932 a definitive work showing the existence of such a 'neutral proton' was carried out by James Chadwick (England, 1891-1974), who named it the neutron. Chadwick was awarded the Nobel Prize for Physics in 1935. The force that was responsible for holding neutrons and protons together inside an atomic nucleus wasn't anything like the two forces of nature known until then. It was way too strong to be of gravitational origin and it clearly was independent of electric charges, for a neutron carried no electric charge. Originally called the nuclear force, the new force went through several name changes — the nuclear force, the strong interaction, the strong force and finally, in something of a compromise, the strong nuclear force. The discovery of the neutron thus marked the beginning of nuclear physics.

1933: The neutrino, a phantom of a particle

In some nuclear radioactive processes, an atomic nucleus transmutes itself into another, lighter one, accompanied by the emission of an electron. When carefully monitored, these processes presented a serious dilemma: the total energy of a system after the transmutation was always slightly less than that before the transmutation, an apparent violation of one of the most sacred rules of nature — the conservation of energy. Something was mysteriously stealing a tiny amount of energy. In a valiant attempt to uphold the principle of energy conservation, Wolfgang Pauli (who proposed the exclusion principle eight years earlier) hypothesized in 1931 a new particle, a particle that had no mass and no electric charge, but existed to provide the necessary energy balance. In 1933 Enrico Fermi (Italy, USA, 1901-1954) named this 'phantom' of a particle the neutrino. The existence of neutrinos was not confirmed until 1955. Both neutrinos and photons share the common characteristics of having neither mass nor charge, but they are entirely different kinds of particles; for one thing, photons are particles of light but neutrinos have no associated radiation in the macroscopic world.

1934: The weak nuclear force

In 1934, Enrico Fermi realized that the type of radioactivity that required invoking neutrinos could be the work of yet another entirely new kind of force. It was affecting neutrons and protons inside a nucleus all right, but in a way completely unexpected from the strong nuclear force. For one thing, the process was too slow and too feeble. Fermi proclaimed the existence of a fourth force and named it the weak nuclear force. He was awarded the Nobel Prize for Physics in 1938 for this work. Despite the similarity in names, the strong and the weak nuclear forces are two entirely different forces — apples and oranges.

1937: Discovery of the muon

In 1937, Carl Anderson (of the positron fame), mined more gold from cosmic ray research, this time by discovering a 'heavy clone' of an electron. The new particle was, in all aspects save one, a carbon copy of the electron; it was identical to the electron in all its behavior but it weighed some 200 times as much. This heavy clone of the electron was named the muon ('mu' as in music).

1948: Quantum electrodynamics

In view of the new physics — relativity and quantum mechanics — a new treatment of electricity and magnetism was called for, a relativistic quantum version of electrodynamics. Richard Feynman (USA, 1918-1988), Julian Schwinger (USA, 1918-1994) and Shinichiro Tomonaga (Japan, 1906-1979), each working independently of the others, rose to the occasion and succeeded in formulating just such a theory in 1948 — the rules for the electromagnetic interplay between electrons and photons. Their work, called quantum electrodynamics or QED for short, represented the most advanced form of a physical theory ever devised by man. The trio shared the Nobel Prize for Physics in 1965.

1955: Neutrinos confirmed

At long last, a quarter of a century after it was first introduced theoretically, the first sighting of the elusive neutrinos was achieved by Frederic Reines (USA, 1918-1998), Clyde Cowan (USA, 1919-1974) and co-workers in 1955 when they identified the telltale tracks using the large flux of neutrinos produced in nuclear reactors in Savannah, Georgia. Forty years later, Reines was recognized for his work with the 1995 Nobel Prize in Physics; Cowan passed away in 1974.

1961: Protons and neutrons may not be elementary, after all

In a series of experiments reminiscent of Rutherford's discovery of the atomic nucleus, physicists at Stanford, led by Robert Hofstadter (USA, 1915-1990), probed the interior of protons and neutrons by shooting a beam of deeply penetrating electrons into them. What they discovered was similar to what Rutherford had originally discovered about atomic nuclei — there were points of concentrated charges inside a proton, which raised the possibility that protons and neutrons themselves may have structures of their own. Hofstadter was awarded one half of the Nobel Prize for Physics in 1961 for this and related works.

1961: The eightfold way

The proliferation in the number of new particles uncovered by high-energy accelerators grew to an 'epidemic' proportion. By 1961, one began to see an emergence of patterns in the groupings of particles with similar attributes, a new 'periodic table of particles,' so to speak. A landmark discovery of systematics was proposed by Murray Gell-Mann (USA, b. 1929) and, independently, by Yuval Ne'eman (Israel, b. 1925): the family of particles fell into distinct groupings, called multiplets, some in groups of eight (octets) and others in groups of ten (decuplet). The Gell-Mann-Ne'eman scheme was christened 'the eightfold way,' a name Gell-Mann borrowed from the teachings of Buddhism. This 'eightfold way' classification paved the way for the next important step, the introduction in 1964 of the idea of quarks. In 1969, Gell-Mann received the Nobel Prize for Physics.

1964: The year of the quark

One thing led to another and the apparent success of grouping particles into families — multiplets — pioneered by Gell-Mann and Ne'eman led to searches for deeper answers. Two people, Gell-Mann and George Zweig (USA, b. 1936) came up with an identical scheme simultaneously and independently: things would make a whole lot of sense if most of the particles — those that reacted to the strong nuclear force — were to be viewed as composites of yet another underlying layer of constituents. In the original proposal, there were three species of such new constituents, named quarks — the up, down and strange quarks (as if the name quark wasn't strange enough). Eventually the number of species of quarks doubled to six.

1964: Why not a charming fourth?

It didn't take long for the need for a fourth quark to come along. By this time, it was well established that there were four types of featherweight particles, collectively called leptons: the electron, the muon and two distinct types of neutrinos. In 1964, Sheldon Glashow (USA, b. 1932) and James Bjorken (USA, b. 1934) proposed a scheme in which there were four types each of leptons and quarks. This did wonders for our understanding of elementary particles. Named the charmed one, this fourth quark rounded off the parallel symmetry between quarks and leptons that remains valid to this day — called the quark-lepton symmetry.

1965: Color charges for quarks

The strong nuclear force, originating from quarks, needed its own characteristic charges, like the electric charges of the electromagnetic force. In 1965 Moo-Young Han (Korea, USA, b. 1934) and Yoichiro Nambu (Japan, USA, b. 1921) introduced the idea of a set of three different 'charges' for quarks. Several years later, this tri-valued property was named 'color' charges by Murray Gell-Mann — the red, green and blue charges. The name has nothing whatever to do with color per se, but the nomenclature stuck and marked the beginning of the quark's tri-color charges.

1974: Discovery of the 'charming' matter

In November 1974, two teams, working independently of each other, simultaneously discovered particles that had all the markings of the charm quark inside. The team at the Brookhaven

Lab in Long Island, led by Samuel C C Ting (USA, b. 1936), called its discovery the J particle; in the West Coast the team at Stanford led by Burton Richter (USA, b. 1931) called it the psi particle. The J/psi particle, as it has come to be called, was the first of many that have since been observed, consistent with the criterion of containing at least one charm quark. Ting and Richter were awarded the Nobel Prize for Physics in 1976.

1975: The tauon: the 'superheavy' electron

Amid a frenzy of activities following the discovery of the J/psi particle, an unexpected bonus was uncovered by yet another team at Stanford, led by Martin Perl (USA, b. 1927). A very heavy relative of the electron, much heavier than the earlier muon, popped into view. Perl shared the 1995 Nobel Prize for Physics with Frederic Reines (USA, b. 1918) who discovered the original neutrino in 1955. A muon, discovered back in 1937, weighs 'only' about 200 times as much as an electron; the new 'super-heavy'— named a tauon or a tau lepton — weighs 3,500 times as much as an electron.

1977: Quark no. 5

When it rains, it pours. A group headed by Leon Lederman (USA, b. 1922) found the evidence for a new particle that helped to establish the need for yet another — this time the fifth — quark. The new particle was dubbed the upsilon, and the fifth quark that is supposed to be its constituent became known as the bottom quark. We now had evidence — albeit indirect — for five types of quarks — up, down, strange, charm and bottom. It launched what turned out to be an 18-year search for quark no. 6, the top quark.

1979: Gluons fall into their place

Gluons are to the strong nuclear force what photons are to the electromagnetic force; the continuous exchange of gluons among quarks defines the strong nuclear force, in exactly the same manner that the electromagnetic force is seen as a continuous exchange of photons. This is the picture originally suggested by Han and Nambu in 1965. Several groups working at the German electron accelerator called the *Deutsches Elektron Synchrotron*, DESY, located in Hamburg, made the first observations of the telltale signature of these gluons in 1979.

1995: The top quark is claimed

Quark no. 6, named the top quark, had eluded detection for 18 years, ever since its existence was inferred from the discovery of no. 5, the bottom quark. Finally, in 1995, a team working at the Fermi National Accelerator Laboratory outside Chicago (FNAL or *Fermilab* for short) first reported sighting it. It has an extremely short lifespan — even by the standards of particle physics where a millionth of a billionth of a second is considered a long time; it breaks up virtually as soon as it is formed. Its existence, like that of so many others, is inferred by its characteristic signature in the pattern of its disintegration The top quark completed the six species of quarks, each in one of the three color charges. Six quarks and six leptons are currently thought of as the end of the line — the ultimate origin of all known matter.

Appendix 2

Powers of Ten

For designating extreme magnitudes, large or small, there is no substitute for the standard scientific notation of the exponents, the powers of ten. From the expanse of the Universe — known to be larger than 10^{28} meters — to the size of quarks inside a proton — estimated to be smaller than 10^{-20} meters — and for every conceivable size and magnitude in between, the exponents in powers of ten provide the measurement that is as precise as it is compact.

The system does have, however, one disadvantage: its sheer compactness can often be misleading with enormous or infinitesimal values alike rendered as virtually meaningless abstract numbers. As an example, a span of time expressed as 31.536×10^7 seconds does not quite deliver the same impact as the words ten years. Human minds, after all, work best in a linear scale, as in 'the scale from 1 to 10,' and the powers of ten are anything but linear. It takes some comparative reflection in our mind before the sense of large exponents, either positive or negative, begin to sink in. Take another example: the annual budget deficit of the United States government is about a billion dollars a day — give or take a few tens of million dollars. That comes to about 42 million dollars an hour or about 700,000 dollars a minute. Expressed as 7×10^5

dollars per minute, the budgetary shortfall becomes abstract: 'one billion dollars a day' and '7×10^5 dollars per minute' are mathematically identical, but they deliver completely different impacts.

Listed below are the standard prefixes up to the powers of 18, in both directions, with their notations and names. For extreme numbers beyond the powers of 18, we revert back to just numerical notations, as in 10^{24} electrons in a cubic centimeter — the average electron destiny of a good metallic conductor. That is about one trillion trillion electrons in a bit of copper about the size of a pea.

The Powers of Ten

Power	Prefix	Notation	Name
10^{18}	exa-	E	quintillion
10^{15}	peta-	P	quadrillion
10^{12}	tera-	T	trillion
10^{9}	giga-	G	billion
10^{6}	mega-	M	million
10^{3}	kilo-	k	thousand
10^{-3}	milli-	m	one-thousandth
10^{-6}	micro-	μ	one-millionth
10^{-9}	nano-	n	one-billionth
10^{-12}	pico-	p	one-trillionth
10^{-15}	femto-	f	one-quadrillionth
10^{-18}	atto-	a	one-quintillionth

[The notation for the prefix micro is the Greek lower case mu, pronounced as in 'music.']

Some prefixes are more familiar than others. The four in the middle — mega, kilo, milli and micro — are in fact in everyday usage, as in megatrends, kilobonuses, milliseconds, or micromanagements. A megahertz, one MHz, is a word that is almost always bound up with computers. It was only a few years ago that we used to be blitzed by such ads as "the power of the new microprocessor operating at the blinding speed of 30 MHz": nowadays, the speed of a personal computer tops out at or near 200 MHz. One hertz, by the way, is the unit of frequency for one cycle per second or, in the case of digital microelectronics, one on-and-off sequence per second. The speed of 100 megahertz means that a microprocessor operates with an internal digital speed of one hundred million on-and-off flashes per second or, equivalently, one instruction is performed by the microprocessor in one hundredth of a millionth of a second.

The next four prefixes— giga, tera, nano and pico— are the benchmarks of today's high technology. Powerful laser beams routinely reach a power output in the range of gigawatts, and some specially-designed ones are known to have achieved — if only for a fraction of a second — an energy output of a few terawatts — a few billion kilowatts. The prefix giga as in gigabits and gigabytes (one byte is equal to eight bits) are fast becoming household words also. 2 Gbit hard disks and CD-ROMs with tens of gigabits of memory are becoming as commonplace as Big Macs and Whoppers. Another cutting-edge 'technoname' involving the prefix giga is the gigaflops, that is, one billion flops. So, what's a flops? One flops stands for one 'floating point operations per second,' a mathematical operation with the freely-moving decimal point. A computational speed of a few gigaflops is more than a quantitative measure; it has become the right to brag of the makers of supercomputers.

At the other end, the prefix nano is finding wider usage these days — nanosecond, nanofabrication and in fact a whole new nanotechnology. The word is, in fact, on the verge of becoming generic, nanoscopy rather than microscopy and nanochips in place of microchips. One nanosecond, a slice of time one billionth of a second short, is time enough for light to cover the distance of about a foot. The unimaginably fast speed of light in empty space — some 180,000 miles per second — finds somewhat down-to-Earth expression in terms of nanoseconds: only one foot per nanosecond.

The four extreme prefixes —exa, peta, femto and atto — are still rarely used. As yet, anyway. Since the frequencies of the visible light spectrum goes from 0.4 to 0.75 petahertz, that is, 4 to 7.5 x 10^{14} Hz, the interval of time between two successive cycles (the inverse of frequency) ranges from 1.3 to 2.5 femtoseconds, that is, a few quadrillionths of a second. In principle, the speed of a 'hyper-supercomputer' in the 21st century could touch the threshold of the petahertz speed, processing data at the speed of a few femtoseconds. We are talking petaflops.

Appendix 3

The Nobel Prizes in Physics

Whenever possible, the dates in which the prize-winning works were accomplished are given in parentheses.

1901 Wilhelm Conrad Röentgen (Germany, 1845–1923) for the discovery of X-rays (1895).

1902 Half each to Hendrik Antoon Lorentz (Netherlands, 1853–1929) and Pieter Zeeman (Netherlands, 1865–1943). Zeeman observed the splitting of spectral lines radiated by excited atoms in magnetic fields (1896). Lorentz had earlier predicted such an effect, thenceforth called the Zeeman effect.

1903 Half to Antoine Henri Becquerel (France, 1852–1908) for his discovery of spontaneous radioactivity (1896), and the other half to Pierre Curie (France, 1859–1905) and Marie Sklodowska Curie (Poland, France, 1867–1934) for their joint researches on radioactivity.

1904 John William Strutt (Lord Raleigh) (England, 1842–1919) for his investigations of the densities of gases and the discovery of argon.

1905 Philipp Eduard Anton Lenard (Hungary, Germany, 1862–1947) for his work on cathode rays (1893).

1906 Joseph John Thomson (England, 1856–1940) for the discovery of electrons (1897).

1907 Albert Abraham Michelson (Germany, USA, 1852–1931) for inventing optical precision instruments (the Michelson-Morley interferometer) and measuring the speed of light (1887).

1908 Gabriel Lippman (France, 1845–1921) for his method of reproducing colors photographically, based on the phenomenon of interference.

1909 Half each to Guglielmo Marconi (Italy, 1874–1937) and Carl Ferdinand Braun (Germany, 1850–1918) for their contributions to the development of wireless telegraphy (1909).

1910 Johannes Diderik van der Waals (Netherlands, 1837–1923) for his work on the equation of state for gases and liquids.

1911 Wilhelm Wien (Germany, 1864–1928) for discovering Wien's law for the blackbody radiation displacement (1893).

1912 Nils Gustaf Dalén (Sweden, 1869–1937) for his invention of automatic regulators for use in conjunction with gas accumulators for illuminating lighthouses.

1913 Heike Kamerlingh-Onnes (Netherlands, 1853–1926) for liquefying helium (1908) and for the discovery of superconductivity (1911).

1914 Max von Laue (Germany, 1879–1960) for establishing the wave nature of X-rays by crystal diffraction (1912).

1915 Half each to William Henry Bragg (England, 1862–1942) and William Lawrence Bragg (Australia, England, 1890–1971), father and son, for their analysis of crystal structure by means of the X-ray diffraction (1913).

1916 The prize was not awarded.

1917 Charles Glover Barkla (England, 1877–1944) for discovering the characteristic X-ray radiation of the elements (1908).

1918 Max Karl Ernst Ludwig Planck (Germany, 1858–1947) in recognition of services rendered to the advancement of physics by his discovery of energy quanta (1900).

1919 Johannes Stark (Germany, 1874–1957) for discovering the Stark effect, the splitting of spectral lines radiated by excited atoms in an intense electric field (1913).

1920 Charles Edouard Guillaume (Switzerland, France, 1861–1938) for services rendered to precise measurements in physics by his discovery of anomalies in nickel steel alloys.

1921 Albert Einstein (Germany, Switzerland, USA, 1879–1955) for services to theoretical physics, and especially for his discovery of the law of the photoelectric effect (1905). [The theory of relativity was not mentioned in the citation!]

1922 Niels Bohr (Denmark, 1885–1962) for his investigations of the structure of atoms, and of the radiation emanating from them (1913).

1923 Robert Andrews Millikan (USA, 1868–1953) for making the first precise determination of the electronic charge (1911) and experimentally verifying Einstein's photoelectric equation (1916).

1924 Karl Manne Georg Siegbahn (Sweden, 1886–1978) for his discoveries and research in the field of X-ray spectroscopy.

1925 Half each to James Franck (Germany, USA, 1882–1964) and Gustav Hertz (Germany, 1887–1975) for their discovery of the Franck-Hertz effect in electron-atom collisions, supporting Bohr's atomic theory (1913).

1926 Jean Baptiste Perrin (France, 1870–1942) for his work on the discontinuous structure of matter (1895).

1927 Half each to Arthur Holley Compton (USA, 1892–1962) for discovering the Compton effect, which showed that a photon has momentum (1923), and to Charles Thomas Rees Wilson (England, 1869–1959) for inventing the expansion cloud chamber, the first major device for detecting charged particles (1911).

1928 Owen Willans Richardson (England, 1879–1959) for his work on the thermionic emission of electrons from hot bodies.

1929 Louis-Victor de Broglie (France, 1892–1987) for having introduced the wave nature of electrons, the beginning of the wave theory of matter (1924).

1930 Sir Chandrasekhara Venkata Raman (India, 1888–1970) for his work on the scattering of light and for the discovery of the Raman effect, the scattering of light by atoms and molecules with a change in wavelength (1928).

1931 The prize was not awarded.

1932 Werner Heisenberg (Germany, 1901–1976) for his contribution in creating quantum mechanics (1925).

1933 Half each to Erwin Schröedinger (Austria, Ireland, 1887–1961) for his contribution in creating quantum mechanics (1926), and to Paul Adrien Maurice Dirac (England, 1902–1984) for developing relativistic quantum mechanics and predicting the existence of an anti-electron (1928).

1934 The prize was not awarded.

1935 Sir James Chadwick (England, 1891–1967) for his discovery of the neutron (1932).

1936 Half each to Victor Franz Hess (Austria, USA, 1883–1964) for discovering the cosmic ray (1910) and to Carl David

Anderson (USA, 1905-1991) for his discovery of the positron, or anti-electron (1932).

1937 Half each to Clinton Joseph Davisson (USA, 1881-1958) and to Sir George Paget Thomson (England, 1892-1975) for their experimental discovery of the diffraction of electrons by crystals, confirming the wave hypothesis of de Broglie (1927).

1938 Enrico Fermi (Italy, USA, 1901-1954) for his demonstrations of the existence of new radioactive elements produced by neutron irradiation.

1939 Ernest Orlando Lawrence (USA, 1901-1958) for inventing the cyclotron (1932).

1940 The prize was not awarded.

1941 The prize was not awarded.

1942 The prize was not awarded.

1943 Otto Stern (Germany, USA, 1888-1969) for discovering the magnetic moments of atoms (1923).

1944 Isidor Isaac Rabi (Austria, USA, 1898-1988) for discovering nuclear magnetic resonance, the basis for today's MRI technology (1935).

1945 Wolfgang Pauli (Austria, Switzerland, 1900-1958) for the discovery of the exclusion principle, also called the Pauli Principle (1924).

1946 Percy Williams Bridgman (USA, 1882-1961) for discoveries made in the field of high-pressure physics.

1947 Sir Edward Victor Appleton (England, 1892-1965) for his investigations of the physics of the upper atmosphere.

1948 Lord Patrick Maynard Stuart Blackett (England, 1897-1974) for his development of the Wilson cloud chamber method and his discoveries therewith in the field of nuclear and particle physics (1933).

1949 Hideki Yukawa (Japan, 1907–1981) for his prediction of the existence of mesons (1935).

1950 Cecil Frank Powell (England, 1903–1969) for his development of the photographic method of studying nuclear processes and his discovery of mesons (1947).

1951 Half each to Sir John Douglas Cockcroft (England, 1897–1967) and to Ernest Thomas Sinton Walton (Ireland, 1903–1995) for their pioneering work on the transmutation of atomic nuclei in an accelerator (1932).

1952 Half each to Felix Bloch (Switzerland, USA, 1905–1983) and to Edward Mills Purcell (USA, b. 1912) for the development of new methods for nuclear magnetic precision measurements (1946).

1953 Frits (Frederik) Zernike (Netherlands, 1888–1966) for his invention of the phase contrast microscope.

1954 Half each to Max Born (Germany, 1882–1970) for his probabilistic interpretation of quantum mechanics (1925) and to Walter Bothe (Germany, 1891-1957) for first identifying the particle later identified as the neutron (1930).

1955 Half each to Willis Eugene Lamb (USA, b. 1913) for his discovery of a small displacement (the Lamb shift) in the fine structure of the hydrogen spectrum (1947), and to Polykarp Kusch (Germany, USA, 1911–1993) for his high-precision determination of the magnetic moment of the electron (1947).

1956 One-third each to William Shockley (England, USA, 1910–1989), Walter Houser Brattain (China, USA, 1902–1987) and

John Bardeen (USA, 1908–1991) for their investigations on semiconductors and their discovery of the transistor (1947).

1957 Half each to Tsung-Dao Lee (China, USA, b. 1926) and Chen-Ning Yang (China, USA, b. 1922) for predicting the non-conservation of parity in beta decay (1956).

1958 One-third each to Pavel Alekseyevich Cherenkov (Russia, 1904–1990) for discovering the Cherenkov radiation (1934), and Igor Yevgenyevich Tamm (Russia, 1895–1971) and Il'ya Mikhailovich Frank (Russia, 1908–1990) for their theoretical interpretation of it (1937).

1959 Half each to Emilio Gino Segrè (Italy, USA, 1905–1989) and Owen Chamberlain (USA, b. 1920) for their discovery of the antiproton (1955).

1960 Donald A Glaser (USA, b. 1926) for the invention of the bubble chamber (1952).

1961 Half each to Robert Hofstadter (USA, 1915–1990) for his pioneering studies of electron scattering in atomic nuclei and for his discoveries concerning the structure of nucleons (1953), and to Rudolf Ludwig Mössbauer (Germany, b. 1929) for discovering the Mossbauer effect of recoilless gamma-ray emission (1958).

1962 Lev Davidovich Landau (Russia, 1908–1968) for his pioneering theories of condensed matter, especially liquid helium.

1963 Half to Eugene P Wigner (Hungary, USA, 1902–1995) for his application of symmetry principles to quantum mechanics, and the other half jointly to Maria Goeppert-Mayer (Germany, USA, 1906–1972) and J Hans D Jensen (Germany, 1907–1973) for developing the shell model of nuclei (1949).

1964 Half to Charles H Townes (USA, b. 1915) and the other half jointly to Nikolai Gennadiyevich Basov (USSR, b. 1922) and Aleksandr Mikhailovich Prokhorov (USSR, b. 1916) for developing masers and lasers (1954).

1965 One-third each to Sinitiro Tomonaga (Japan, 1906–1979), Richard P Feynman (USA, 1918–1988) and Julian Schwinger (USA, 1918–1994) for developing the theory of quantum electrodynamics (1948).

1966 Alfred Kastler (France, 1902–1984) for his optical methods for studying atomic energy levels.

1967 Hans Albrecht Bethe (Germany, USA, b. 1906) for his contributions to the theory of nuclear reactions, especially his discoveries concerning the energy production in stars (1939).

1968 Luis W Alvarez (USA, 1911–1988) for discovering a large number of resonant states of elementary particles.

1969 Murray Gell-Mann (USA, b. 1929) for his contributions concerning the classification of elementary particles and their interactions (1961).

1970 Half each to Hannes Alfvén (Sweden, 1908–1995) for his fundamental work in magneto-hydrodynamics, and to Louis Néel (France, b. 1904) for his discoveries in antiferromagnetism and ferromagnetism (1930s).

1971 Dennis Gabor (Hungary, England, 1900–1979) for his invention of holography (1947).

1972 One-third each to John Bardeen (USA, 1908–1991), Leon N Cooper (USA, b. 1930) and John Robert Schrieffer (USA, b. 1931) for their theory of superconductivity, usually called the BCS theory (1957). This was Bardeen's second Nobel Prize in Physics.

1973 One half was shared between Leo Esaki (Japan, USA, b. 1925) and Ivar Giaever (Norway, USA, b. 1929) for their experimental discoveries regarding tunneling phenomena in semiconductors and superconductors, respectively, and the other half went to Brian D Josephson (England, b. 1940) for predicting the Josephson effect.

1974 Half each to Sir Martin Ryle (England, 1918–1984) for developing radio interferometry, and Anthony Hewish (England, b. 1924) for discovering pulsars (1967).

1975 One-third each to Aage Bohr (Denmark, b. 1922), Ben Mottelson (Denmark, b. 1926) and James Rainwater (USA, 1917–1986) for the development of the theory of the structure of atomic nuclei based on a non-spherical nuclear core (1950–1952).

1976 Half each to Burton Richter (USA, b. 1931) and Samuel C Ting (USA, b. 1936) for their independent discovery of the J/psi particles, the first particles made of the charm quarks (1974).

1977 One-third each to Philip W Anderson (USA, b. 1923), Sir Neville F Mott (England, b. 1905) and John H Van Vleck (USA, 1899–1980) for their contributions in the study of the electronic structure of magnetic systems.

1978 Half to Pyotr Leonidovich Kapitsa (Russia, 1894–1984) for his studies of liquid helium, and the other half jointly to Arno A Penzias (Germany, USA, b. 1933) and Robert W Wilson (USA, b. 1936) for their discovery of cosmic microwave background radiation (1965).

1979 One-third each to Sheldon L Glashow (USA, b. 1932), Abdus Salam (Pakistan, b. 1926) and Steven Weinberg (USA, b. 1933) for developing the theory that unified the weak nuclear and electromagnetic forces.

1980 Half each to James W Cronin (USA, b. 1931) and Val L Fitch (USA, b. 1923) for the discovery of the time-reversal invariance violation in the decay of neutral K-mesons (1964).

1981 Half to Nicolaas Bloembergen (Netherlands, USA, b. 1920) and Arthur L Schawlow (USA, b. 1921) for their contributions to the development of laser spectroscopy, and the other half to Kai M Siegbahn (Sweden, b. 1918) for his contribution to the development of high-resolution electron spectroscopy.

1982 Kenneth G Wilson (USA, b. 1936) for his theory of critical phenomena in connection with phase transitions.

1983 Half to Subramanyan Chandrasekhar (India, USA, 1910–1995) for his theoretical studies of the physical processes of importance to the structure and evolution of the stars (1930), and the other half to William A Fowler (USA, 1911–1995) for his theoretical studies of astrophysical nucleosynthesis.

1984 Half to Carlo Rubbia (Italy, b. 1934) for discovering the W and Z particles, and the other half to Simon Van der Meer (Netherlands, b. 1925) for developing the experimental method that allowed the discovery.

1985 Klaus von Klitzing (Germany, b. 1943) for the discovery of the quantized Hall effect (1980).

1986 Half to Ernst Ruska (Germany, 1906-1988) for inventing the electron microscope (1931), and the other half jointly to Gerd Binnig (Germany, b. 1947) and Heinrich Rohrer (Switzerland, b. 1933) for their invention of the scanning tunneling microscope (1981).

1987 Jointly to J Georg Bednorz (Germany, b. 1950) and K Alexander Müeller (Switzerland, b. 1927) for their discovery of high-temperature superconductors (1986).

Appendix 3: The Nobel Prizes in Physics

1988 One-third each to Leon M Lederman (USA, b. 1922), Melvin Schwartz (USA, b. 1932) and Jack Steinberger (Germany, USA, b. 1921) for their work leading to the discovery of the muon neutrino.

1989 Half to Norman F Ramsey (USA, b. 1915) for his contributions to atomic clocks, and the other half jointly to Hans G Dehmelt (Germany, USA, b. 1922) and Wolfgang Paul (Germany, 1913–1993) for the development of the ion trap technique.

1990 Awarded jointly to Jerome I Friedman (USA, b. 1930), Henry W Kendall (USA, b. 1926), and Richard E Taylor (Canada, b. 1929) for their pioneering investigations concerning deep inelastic scattering of electrons on protons and bound neutrons, which have been of importance for the quarks model of particle physics.

1991 Pierre-Gilles de Gennes (France, b. 1932) for his contributions to the study of complex forms of matter, in particular to liquid crystals and polymers.

1992 Georges Charpak (Poland, France, b. 1924) for developing the multiwire proportional chamber for detecting elementary particles.

1993 Jointly to Russel Hulse (USA, b. 1950) and Joseph H Taylor (USA, b. 1941) for their discovery of a new type of pulsar.

1994 Half each to Clifford Schull (USA, b. 1915) and Bertram Brockhouse (Canada, b. 1918) for developing the neutron spectroscopy.

1995 Half each to Frederick Reines (USA, 1918–1998) for the detection of the neutrino, and to Martin Perl (USA, b. 1927) for the discovery of the tau lepton.

1996 Awarded jointly to David M Lee (USA, b. 1931), Douglas D Osheroff (USA, b. 1945), and Robert C Richardson (USA, b. 1937) for their discovery of superfluidity in helium-3.

1997 Awarded jointly to Steven Chu (USA, b. 1948), Claude Cohen-Tannoudji (France, b. 1933), and William D Phillips (USA, b. 1948) for the development of methods to cool and trap atoms with laser light.

1998 Awarded jointly to Robert B Laughlin (USA, b. 1950), Horst L Störmer (Germany, b. 1949), and Daniel C Tsui (China, USA, b. 1939) for the discovery of a new form of quantum fluid.